RICE BIOTECHNOLOGY:
IMPROVING YIELD, STRESS TOLERANCE AND GRAIN QUALITY

The Novartis Foundation is an international scientific and educational charity (UK Registered Charity No. 313574). Known until September 1997 as the Ciba Foundation, it was established in 1947 by the CIBA company of Basle, which merged with Sandoz in 1996, to form Novartis. The Foundation operates independently in London under English trust law. It was formally opened on 22 June 1949.

The Foundation promotes the study and general knowledge of science and in particular encourages international co-operation in scientific research. To this end, it organizes internationally acclaimed meetings (typically eight symposia and allied open meetings and 15–20 discussion meetings each year) and publishes eight books per year featuring the presented papers and discussions from the symposia. Although primarily an operational rather than a grant-making foundation, it awards bursaries to young scientists to attend the symposia and afterwards work with one of the other participants.

The Foundation's headquarters at 41 Portland Place, London W1N 4BN, provide library facilities, open to graduates in science and allied disciplines. Media relations are fostered by regular press conferences and by articles prepared by the Foundation's Science Writer in Residence. The Foundation offers accommodation and meeting facilities to visiting scientists and their societies.

Information on all Foundation activities can be found at
http://www.novartisfound.org.uk

Novartis Foundation Symposium 236

RICE BIOTECHNOLOGY:
IMPROVING YIELD, STRESS TOLERANCE AND GRAIN QUALITY

2001

JOHN WILEY & SONS, LTD

Chichester · New York · Weinheim · Brisbane · Singapore · Toronto

Other Wiley Editorial Offices

John Wiley & Sons, Inc., 605 Third Avenue,
New York, NY 10158-0012, USA

WILEY-VCH Verlag GmbH, Pappelallee 3,
D-69469 Weinheim, Germany

Jacaranda Wiley Ltd, 33 Park Road, Milton,
Queensland 4064, Australia

John Wiley & Sons (Asia) Pte Ltd, 2 Clementi Loop #02-01,
Jin Xing Distripark, Singapore 129809

John Wiley & Sons (Canada) Ltd, 22 Worcester Road,
Rexdale, Ontario M9W 1L1, Canada

Novartis Foundation Symposium 236
x+261 pages, 43 figures, 10 tables

Library of Congress Cataloging-in-Publication Data

Rice biotechnology: improving yield, stress tolerance and grain quality / [edited by
Jamie Goode, Derek Chadwick].
        p. cm. – (Novartis Foundation symposium ; 236)
     Includes bibliographical references (p.).
     ISBN 0-471-49661-8 (alk. paper)
     1. Rice–Biotechnology–Congresses. I. Goode, Jamie. II. Chadwick, Derek. III.
Symposium on Rice Biotechnology: Improving Yield, Stress Tolerance, and Grain
Quality (2000 : Laguna, Philippines)   IV. Series.
SB191.R5 R4456 2001
633.1′8233–dc21   2001017603

British Library Cataloguing in Publication Data

A catalogue record for this book is available from the British Library

ISBN 0 471 49661 8

Typeset in $10\frac{1}{2}$ on $12\frac{1}{2}$ pt Garamond by Dobbie Typesetting Limited, Tavistock, Devon.
Printed and bound in Great Britain by Biddles Ltd, Guildford and King's Lynn.
This book is printed on acid-free paper responsibly manufactured from sustainable forestry,
in which at least two trees are planted for each one used for paper production.

Contents

Participants

John Bennett   Division of Plant Breeding, Genetics and Biochemistry, International Rice Research Institute, DAPO Box 7777, Metro Manila, Philippines

Peter D. Beyer   Institut fur Biologie II, Zelbiologie, Universitat Freiburg, Schanzlestrasse 1, D-79104, Freiburg, Germany

Howarth Bouis   International Food Policy Research Institute, 2033 K St NW, Washington, DC 20006, USA

Xinnian Dong   Duke University, DCMB Department of Botany, LSRC Building, Research Drive, Durham, NC 277706-1000, USA

Malcolm Elliott   Norman Borlaug Institute for Plant Science Research, De Montfort University, Scraptoft, Leicester LE7 9SU, UK

Mike Gale   John Innes Centre, Norwich Research Park, Colney, Norwich, NR4 7UH, UK

Stephen A. Goff   Novartis Agricultural Discovery Institute, 3115 Merryfield Row Suite 100, San Diego, CA 92121-1125, USA

Robin D. Graham   Department of Plant Science, University of Adelaide, Waite Campus, Glen Osmond, South Australia 5064, Australia

Peter Horton   Department of Molecular Biology and Biotechnology, University of Sheffield, Western Bank, Sheffield, S10 2TN, UK

Mahabub Hossain   Social Sciences Division, International Rice Research Institute, DAPO Box 7777, Metro Manila, Philippines

Gurdev S. Khush (*Chair*)   Division of Plant Breeding, Genetics and Biochemistry, International Rice Research Institute, DAPO Box 7777, Metro Manila, Philippines

Maurice Ku   Department of Botany, Washington State University, Pullman, WA 99164-4238, USA

Jan E. Leach   Department of Plant Pathology, 4024 Throckmorton Plant Sciences Center, Kansas State University, Manhattan, KS 66506-5502, USA

Hei Leung   International Rice Research Institute, DAPO Box 7777, Metro Manila, Philippines

Zhikang Li   International Rice Research Institute, DAPO Box 7777, Metro Manila, Philippines

Takashi Matsumoto   National Institute of Agrobiological Resources, 2-1-2 Kannondai, Tsukuba, Ibaraki 305-8602, Japan

Barbara Mazur   DuPont Agricultural Products, P.O. Box 80402, Wilmington, DE 19880-0402, USA

Graham McLaren   Department of Biometrics, International Rice Research Institute, DAPO Box 7777, Metro Manila, Philippines

David Nevill   Novartis Crop Protection Munchwilen AG, Postfach, CH-4332 Stein, WST-540.1.43, Switzerland

Thomas W. Okita   Washington State University, Institute of Biological Chemistry, PO Box 646340, Pullman, WA 99164-6340, USA

Jane Parker   The Sainsbury Laborary, John Innes Centre, Norwich Research Park, Colney, Norwich NR4 7UH, UK

Gernot Presting   Torrey Mesa Research Institute, 3115 Merryfield Row, San Diego, CA 92121-1125, USA

Tony Pryor   Division of Plant Industry, CSIRO, P.O. Box 1600, Canberra, ACT 2601, Australia

John Salmeron   Novartis Agribusiness Biotechnology Research Inc, 3054 Cornwallis Road, PO Box 12257, Research Triangle Park, NC 27709, USA

Guo-Liang Wang   Department of Plant Pathology, Ohio State University, Kottman Hall, 2021 Coffey Road, Columbus, Ohio 43210-1087, USA

Ren Wang   International Rice Research Institute, DAPO Box 7777, Metro
Manila, Philippines

Kazuko Yamaguchi-Shinozaki   Biological Resources Division, Japan
International Research Center for Agricultural Sciences (JIRCAS), Ministry of
Agriculture, Forestry and Fisheries, 1-2 Oowashi, Tsukuba, Ibaraki 305-8686,
Japan

Opening address: The challenge to feed the World's poor

Mahabub Hossain

International Rice Research Institute, DAPO Box 7777, Metro Manila, Philippines

Food is the most basic human need. At low levels of income the utmost concern for human beings is to meet the energy needs to overcome hunger, and cereals provide the cheapest source of energy. Thus per capita intake of the staple foods — rice, wheat, maize and tubers — is usually high at low levels of income and increases further with rising incomes. Cereal intake, however, starts declining when the basic energy needs are met (at a middle income level), when people can afford to have a more diversified diet that provides balanced nutrition with adequate consumption of vegetables, meat and livestock products that are rich in protein, vitamins and micronutrients. But as the demand for livestock products increases with economic prosperity, so does the indirect demand for some cereals, such as maize and other coarse grains that are used as livestock feed. The decline in per capita consumption of cereals for human food is over-compensated by the increase in the per capita demand for cereals as livestock feed (Alexandratos 1995). The increase in the per capita demand for cereals therefore goes hand in hand with economic prosperity.

Despite the closing down of the land frontier, particularly in the densely settled countries in Asia, the World has done remarkably well in meeting the food needs of its fast growing population over the last four decades (Pinstrup-Andersen 1994, Pinstrup-Andersen & Lorch 1997). The phenomenal increase in agricultural productivity, made possible by the adoption of genetically improved varieties, has enabled many countries to increase food supplies faster than demand, to nullify the dire predictions of food insecurity and famine (Paddock & Paddock 1967, Brown 1974, Eckholm 1976). Technological progress contributed to a decline in unit cost of production which enabled farmers to share the benefits with consumers by offering food to them at affordable prices. The long term decline in the real prices of food was the major factor contributing to poverty alleviation even in agroecosystems that were bypassed by the 'green revolution' (Conway 1998, David & Otsuka 1994).

But much remains to be done. Although enough food is being produced today at the global level, the unequal access to food due to lack of purchasing capacity for

1

the low-income people of the poorest countries contributes to hunger and undernourishment in nearly 800 million people. Nearly two-thirds of them live in Asia where rice is the dominant food staple, despite remarkable economic progress made over the last four decades. The challenge to cutting edge science today is whether a new 'doubly green revolution' — one that is environmentally sustainable as well as yield-increasing — can meet the food needs of the low-income poverty-stricken countries that will not reach a stationary population till the end of the next century.

Population growth remains the major threat to food insecurity

The most important factor determining the demand for cereals is the continued increase in the number of mouths to be fed. The world population has tripled over the last 50 years, from 2.5 billion in 1950 to 6.1 billion in 2000, and is not expected to stabilize before 2100, when the number may reach between 9.4 (with faster progress in fertility decline) to 11 billion (with standard fertility decline) (World Bank 1996, United Nations 1998). Most of the increase in population will be in the developing countries. Over the next 25 years the world population is projected to increase by 1.95 billion; nearly 31% of the increase will come from south Asia and another 28% from sub-Saharan Africa (Table 1). These are the regions in the world where poverty and hunger are widespread and the per capita cereal consumption is less than half of that for the developed countries.

The developed countries may not need further increase in cereal production to meet their internal demand. Most of these countries have reached a stationary population and many, particularly in Europe, will experience an absolute decline in population very soon. Only in North America and Australia is population still growing, mostly due to in-migration of people from the low-income countries. In Europe, North America and Japan, the per capita consumption of cereals has also started declining because of the sluggish internal demand for livestock products and growing consumers' preference for low-calorie diets with dominance of vegetables and fruits. With further increase in yields these countries are now facing problems of disposal of surplus and a downward pressure on the prices of cereals in the world market that dampens farmers' incentives to grow more food. Many countries have adopted a policy of direct payment to farmers to reduce land under cereal crops to lower production and maintain prices. In Japan, for example, the area under rice has declined from 3.3 million ha in 1960 to about 2.0 million ha in 1999, in response to the declining domestic demand for rice. Many farmers in developed countries now find it profitable to go for organic farming, with reduced crop yield but higher market price because of the affluent consumers' preference for organic food.

TABLE 1   Population growth and increase in food requirements for different regions in the World, 2000–2025

Regions	Population (billions)		Projected per capita consumption of cereals (kg/yr)		Food grain requirements (million tons)		Percentage increase
	2000	2025	2000	2025	2000	2025	2000–2025
East Asia	1.48	1.70	284	332	420	564	34
South-central Asia	1.50	2.10	167	187	250	392	57
Southeast Asia	0.52	0.69	210	242	109	167	53
Western Asia & North Africa	0.36	0.55	405	469	146	258	77
Sub-Saharan Africa	0.65	1.20	138	156	90	187	108
Latin America	0.52	0.69	273	301	142	208	46
Developing Countries	4.90	6.82	258	280	1265	1910	51
Developed Countries	1.19	1.22	626	680	745	830	11

Source: Estimate from Rosegrant et al (1995), Alexandratos et al (1995) and United Nations (1998).

The situation in the developing countries is completely opposite because of the continuing high growth of population (Fig. 1). It will take a long time for many developing countries to reach the present level of per capita cereal consumption of the developed world (see Table 1). So, it is expected that the cereal intake will follow an increasing trend until these countries reach the middle-income stage, and the rate of increase will depend on the growth of incomes. Ironically, in the poverty-stricken regions, such as in sub-Saharan Africa and South Asia, the per capita consumption is expected to increase and the population will also grow faster, and thus it will take longer to reach the stage of stationary population. Within the next 25 years the food requirement is projected to double in sub-Saharan Africa and grow by 50–75% in other regions of the developing world. Only in East Asia is the growth in demand for cereals expected to slacken (Table 1).

Producing more food with less pressure on natural resources

Increasing cereal production to meet the massive increase in food grain requirement in the developing world will not be an easy task without continuing efforts towards crop improvement, particularly for rice (South and Southeast

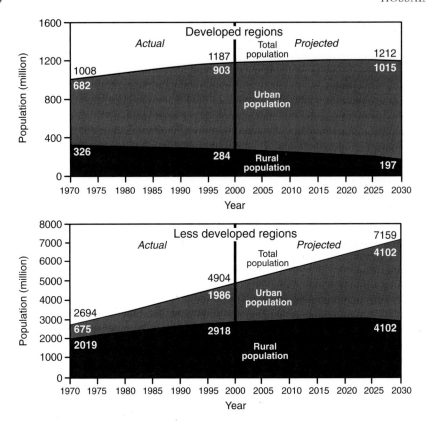

FIG. 1. Population growth and urbanization: recent trends (1970–2000) and future prospects (2000–2030). Source: United Nations (1998).

Asia), maize (Sub-Saharan Africa, and Southeast Asia) and root crops (Africa). In Asia, the land frontier has long been exhausted, and the increase in cropped area in the past has come from more intensive utilization of land for raising two to three crops per year which has put pressure on sustaining the natural resource base (Brown 1996, Pingali et al 1997). With growing urbanization and industrialization, some of the fertile agricultural land has gradually been diverted to meet the demand for housing, factories and roads. The perception of abundance of water, another key natural resource, has been changing even for humid tropics and subtropics, with competing demand arising from the growth of population, urbanization and industrialization. In Africa, the abundant land resources cannot be brought under high-productive agriculture because of low soil fertility, the scarcity of water resources and the high-cost of water resource development projects. Environmental concern regarding the adverse effects of irrigation and flood control projects on water logging, build-up of soil salinity, fish production and the quality of ground water, has been growing and further withdrawal of fresh water resources for additional food production is no longer an easy choice. Land covered by crops year round with increasing intensity of cropping provides an excellent habitat for pests. With the movement from a less-intensive low-yield production system to a more-intensive high-yield system, pest pressure has been growing, and so has the use of agrochemicals harmful to human health and the environment (Rola & Pingali 1993). Even labour is getting scarce, and the wage rate is rising faster than food grain prices with opportunities for more remunerative employment created in the fast-growing non-farm sectors of the economy (Hossain 1996). The challenge to countries that are poorly endowed with natural resources is therefore how to produce more foodgrains from less land, with less water, less labour and less harmful agrochemicals.

The developed countries—North America, Europe and Australia—have abundant natural resources to produce enough food for the world. The pressure on natural resources in the food-deficit countries would be less if the developed countries exploited their unused capacity to produce surplus food to meet the deficit in countries with limited natural resources.

The international trade of cereals—the movement of surplus grains from developed to developing countries through the market mechanism—may be only a part of the solution of the mismatch of the demand–supply balances between the developed and the developing regions of the world. Farmers in developed countries may expect an expansion of market for wheat and corn as livestock feed but this is mainly in the middle-income countries of the developing world that can afford to pay for such transactions. But for staple food, such as rice in Asia and West Africa, and maize and root crops in other parts of the developing world, the growth of the export market for staple food grown in developed countries will remain limited. A large part of the future demand for human food will originate from South-Asia and sub-Saharan Africa. But these countries do not have foreign exchange earnings to pay for commercial transactions for staple food.

Also at low-levels of income, the production of staple food is the major source of employment and incomes for the people. The growth in productivity and production of staple food within the country is considered by policy makers a high priority strategy considering the socio-political compulsion of generating employment and income for the poor farm producers and consumers. So, meeting the demand–supply gap through further improvement of the food grain crops within the national borders, rather than through trade, may be considered an appropriate strategy for addressing the problem of food insecurity and poverty in the developing world.

Indeed, there is a large yield gap between developed and developing countries for both rice and maize, which are the dominant food staple for Asia and sub-

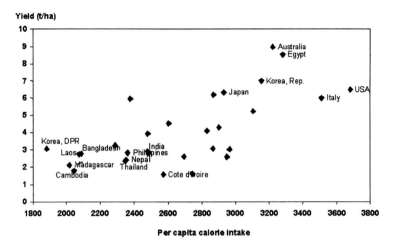

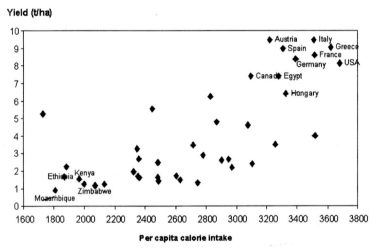

FIG. 2.   Yield gaps for rice and maize across countries at different levels of per capita calorie intake, 1997–1998 (Source: FAO electronic database).

Saharan Africa, which could be narrowed through genetic enhancement research appropriate for the specific agro-ecosystems that characterize these regions. The energy intake of people is strongly associated with the yield of the staple grains (Fig. 2), which suggests that an increase in yield would contribute to amelioration of hunger. Technological progress at the country level is thus still of paramount importance in our struggle for feeding the world.

Genetic enhancement in rice: achievements and limitations

The major achievements in genetic advancement in rice have so far been in: (a) a substantial shift in yield potential through improving nitrogen-responsiveness of rice plants and increasing harvest index through introducing the dwarfing genes; (b) incorporation of host-plant resistance against major insects and diseases; and (c) reduction in the crop maturity period from over 150 days to below 100 days (Khush 1995). These advances helped farmers save land, the scarcest natural resource in Asia, increase cropping intensity (for irrigated land) and crop diversification (for rainfed land), and reduce farmers' dependence on harmful agrochemicals. In order to reap full benefits of these technologies farmers had to make associated investments for water control (flood control, drainage and irrigation) and in chemical fertilizers. So it is the farmers with better endowment of resources who had an advantage in adopting these technologies. Only limited progress has so far been made in improving grain quality, and developing tolerance against problem soils.

Large parts of Asia, particularly the uplands and the rainfed lowlands, are yet to benefit from these technological innovations because the new varieties are not adaptable to these ecosystems. Scientists have had limited success in developing appropriate high-yielding rices that can adapt to the floods, droughts, temporary submergence caused by heavy rains and poor drainage, and problem soils that characterize these ecosystems. Available improved varieties may do well in normal years but perform poorly compared to the traditional land races if there is a prolonged drought or sudden submergence due to the vagaries of monsoons. If the rainy season is short, or the rainfall is unreliable, and therefore the risk of cultivation is high, the risk-averse small farmers will grow traditional varieties and use fertilizers in small amounts. So the diffusion of the improved varieties remained limited to favourable environments.

Figure 3 illustrates that the green revolution in rice cultivation has largely benefited the irrigated regions that already had high yield before the green revolution. The yield gap between the favourable irrigated and the unfavourable rainfed environments has grown over time because of the vastly poor performance of the latter with regard to increase in yields. It is the regions with predominantly rainfed rice ecosystems (Eastern India, Myanmar, Northeast Thailand, Cambodia, Laos, Visayas and Mindanao in the Philippines, and highlands in Vietnam and Indonesia) where poverty, food insecurity and malnutrition are still widespread. The rainfed ecosystems still account for 55% of the rice land and 45% of the rice harvested area, although they account for only 25% of the rice production in Asia because of the low yield. If the utilization of the cutting edge science in rice research is to make impact on alleviation of poverty, researchers must focus on addressing the problems of these unfavourable rice-growing environments.

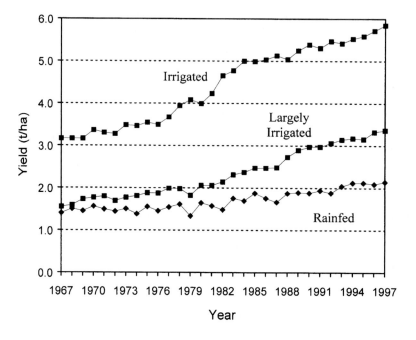

	Average yield (t/ha)			Growth rate (%/yr)	
Ecosystem	1967–69	1984–86	1995–97	1967–85	1985–97
Irrigated	3.2	5.0	5.7	2.7 (0.2)	1.3 (0.1)
Largely Irrigated	1.6	2.4	3.3	2.2 (0.2)	2.7 (0.2)
Rainfed	1.5	1.8	2.1	0.9 (0.3)	1.8 (0.3)

Note: The figure within parentheses is the standard error of the estimated growth rate.
Sources: IRRI World Rice Statistics and FAO (1999).

FIG. 3.   Trends in rice yield for irrigated and rainfed ecosystems, Asia, 1967–1997.

Conclusions

Recent advances in science, particularly in the field of molecular biology, have the potential to contribute to further improvement in food security and poverty alleviation if their application focuses on addressing the problems of unfavourable agro-ecosystems. Structural and functional genomics can help discover genes that confer on low-yielding land races the capacity to withstand prolonged droughts and submergence and to survive on problem soils. Rice breeders can expedite breeding and improve breeding efficiency with marker-assisted selection techniques for genetic enhancement that incorporate abiotic stress tolerance into high yields. Genetic engineering has already demonstrated the usefulness of introducing valuable traits in rice through transformation, traits that conventional plant breeding approaches have not been able to introduce. Incorporating traits of C4 plants that have higher photosynthesis efficiency into rice can help reduce poor farmers' dependence on chemical fertilizers. Already potential has been demonstrated in increasing the micronutrient content in rice by transforming genes from other species into japonica rice, which is consumed by high-income consumers. This innovation can make important contribution in reducing malnutrition for the urban and rural poor if this genetic enhancement is targeted for the coarse, low-price indica varieties that the poor consume.

Unfortunately, the current application of molecular biology tools in rice research mostly focuses on traits related to pest resistance (Hossain et al 2000). These traits will benefit farmers who operate in irrigated environments with high yields, in countries where the demand for rice has been slackening recently. Also, substantial research capacity for addressing these problems through conventional approaches exists in national agricultural research systems (NARS) in Asia due to the efforts of human capital development through decades of training and collaborative research with NARS, the International Rice Research Institute and the Universities in the developed countries. Biotechnology research to address biotic stress tolerance will merely replace the existing capacity and hence will add only marginal value. To benefit the poor with the application of cutting-edge science, rice research must be targeted to address the problems of abiotic stresses and human nutrition, problems that are found predominantly in unfavourable rice-growing environments and in regions with marginal lands. It is here where substantial value can be added with the biotechnology research.

References

Alexandratos N 1995 World agriculture towards 2010. Wiley, Chichester

Brown LR 1974 By bread alone. Paeger Publishers, New York

Brown LR 1996 Tough choices: facing the challenge of food scarcity. WW Norton, New York

Conway G 1998 The doubly green revolution: food for all in the twenty-first century. Cornell University Press, Ithaca, NY

David C, Otsuka K 1994 Modern rice technology and income distribution in Asia. Lynne Rienner, Boulder, CO

Eckholm EP 1976 Losing gound. WW Norton, New York

Hossain M 1996 Economic prosperity in Asia: implications for rice research. In: Khush GS (ed) Rice genetics III. Proceedings of the Third International Rice Genetics Symposium, Manila, 16–20 October 1995. International Rice Research Institute, Philippines, p 3–16

Hossain M, Bennett J, Datta S, Leung H, Khush G 2000 Biotechnology research in rice for Asia: priorities, focus and directions. In: Qaim M, Krattiger A, von Braun J (eds) Agricultural biotechnology in developing countries: towards optimizing the benefits for the poor. Kluwer Academic Publishers, Norwell, MA p 99–120

Khush GS 1995 Breaking the yield frontier of rice. GeoJournal 35:286–298

Paddock W, Paddock P 1967 Time of famine. Little, Brown, Boston, MA

Pingali PL, Hossain M, Gerpacio RW 1997 Asian rice bowls: the returning crisis? CAB International, Wallingford

Pinstrup-Andersen P 1994 World food trends and future food security. International Food Policy Research Institute, Washington, DC (Food policy statement 18)

Pinstrup-Andersen P, Lorch RP 1997 Food security: a global perspective. In: Peters GH, von Braun J (eds) Food security, diversification, and resource management: refocusing the role of agriculture? proceedings of the twenty-third international conference of agricultural economists, Sacramento, CA, August 1997. Ashgate Publishing, Vermont, p 51–76

Rola AC, Pingali PL 1993 Pesticides, rice productivity and farmers' health: an economic assessment. World Resources Institute, Washington, D.C. and International Rice Research Institute, Manila, Philippines

Rosegrant MW, Agcaoili-Sombilla M, Perez ND 1995 Global food projections to 2020: implications for investment. International Food Policy Research Institute, Washington, DC (Discussion paper 5)

United Nations 1998 World urbanization prospects: the 1996 revision. Population Division, Department of Economic and Social Affairs of the UN Secretariat, New York

World Bank 1996 World development report: from plan to market. Oxford University Press, Oxford

Chair's introduction

Gurdev Khush

Division of Plant Breeding, Genetics and Biochemistry, International Rice Research Institute, DAPO Box 7777, Metro Manila, Philippines

Of the food crops that feed the world, the three cereals, maize, rice and wheat provide 49% of the calories consumed by the human population. And of the total calories consumed globally, rice represents 23%, wheat 17% and maize 9%. The overwhelming importance of rice in feeding the world is clear from these figures. 92% of the world's rice is produced in Asia, where it provides between 30–76% of the calories consumed by humans.

Much progress has been made in increasing rice production: it has risen from 257 million tons in 1966, when the first of the new varieties was released, to almost 600 million tons today. This increase has so far kept up with the corresponding increase in population. In spite of this, there are many people in Asia who are still malnourished, and many of the world's food-insecure live in countries where rice is the primary crop. The world population is increasing by the rate of 1.7–1.8%, adding some 80 million people each year, with the majority of this growth occurring in the rice-consuming nations.

To meet this increased demand, we must produce 50% more rice during the next 25–30 years. This increased production must come from better technology, because there is no more land to open up for rice cultivation: in fact, since 1980 there has been very little increase in the area planted to rice. Over this period we will have to cope with less water for agriculture, less labour, less land (because of demand for land for industrialization and infrastructure development) and fewer chemical inputs. We will therefore have to produce this extra food with these constraints.

There are two key approaches that we will address at this meeting.

(a) *The development of varieties with increased yield potential.* The maximum yield potential of rice is currently 10 tons/ha. We have to develop new varieties with 20–25% higher yield in the next 15 years.
(b) *Closing the yield gap.* Although the maximum yield is 10 tons/ha, under irrigated conditions farmers still average only about 5 tons/ha. How can we get this up to 7–8 tons/ha? This can be done by developing varieties with durable disease and pest resistance. Also, under rainfed conditions — about

40% of the rice area — there are many abiotic stresses such as drought, salinity and excess water. We need varieties with tolerance to these stresses. To give some perspective, in rice, 9% crop loss is caused by disease and 27.3% by insect pests. If we can reduce these, we can produce a lot more rice.

The focus of this symposium will be discussing the technologies for developing improved rice varieties through advanced genetic techniques. I look forward to our discussions on these topics.

A framework for sequencing the rice genome

Gernot G. Presting[1], Muhammad A. Budiman, Todd Wood, Yeisoo Yu, Hye-Ran Kim, Jose Luis Goicoechea, Eric Fang, Barbara Blackman, Jiming Jiang, Sung-Sick Woo, Ralph A. Dean, David Frisch and Rod A. Wing

Clemson University Genomics Institute, 100 Jordan Hall, Clemson University, Clemson, SC 2963-5708, USA

Abstract. Rice is an important food crop and a model plant for other cereal genomes. The Clemson University Genomics Institute framework project, begun two years ago in anticipation of the now ongoing international effort to sequence the rice genome, is nearing completion. Two bacterial artificial chromosome (BAC) libraries have been constructed from the *Oryza sativa* cultivar Nipponbare. Over 100 000 BAC end sequences have been generated from these libraries and, at a current total of 28 Mbp, represent 6.5% of the total rice genome sequence. This sequence information has allowed us to draw first conclusions about unique and redundant rice genomic sequences. In addition, more than 60 000 clones (19 genome equivalents) have been successfully fingerprinted and assembled into contigs using *FPC* software. Many of these contigs have been anchored to the rice chromosomes using a variety of techniques. Hybridization experiments have shown these contigs to be very robust. Contig assembly and hybridization experiments have revealed some surprising insights into the organization of the rice genome, which will have significant repercussions for the sequencing effort. Integration of BAC end sequence data with anchored contig information has provided unexpected revelations on sequence organization at the chromosomal level.

2001 Rice biotechnology: improving yield, stress tolerance and grain quality. Wiley, Chichester (Novartis Foundation Symposium 236) p 13–27

Rice, the world's most important food crop and a model for monocot genomics, has been selected by the international community to be completely sequenced. The objective of the Clemson University Genomics Institute (CUGI) Rice Genome Framework Project, initiated in 1998, is to provide the foundation for efficient sequencing of the rice genome using the sequence-tagged connector (STC) approach (Mahairas et al 1999).

---

[1]Present address: Torrey Mesa Research Institute, 3115 Merryfield Row, San Diego, CA 92121, USA.

The specific objectives were to (a) construct two large insert rice bacterial artificial chromosome (BAC) libraries using different restriction enzymes, (b) end sequence all clones of both libraries, and (c) assemble a physical map by fingerprinting 60 000 BAC clones and anchoring the resulting contigs to the genetic map using molecular markers.

Rice BAC library construction

Two libraries of the rice (*Oryza sativa* ssp. *japonica*) cultivar Nipponbare were constructed using *Hin*dIII and *Eco*RI as the cloning enzymes (M. A. Budiman, M. Luo, J. P. Tomkins, H. Kim & R. A. Wing, unpublished results). The average insert sizes of the two libraries are 130 kb and 120 kb, respectively, and their combined genome coverage exceeds 25×. Hybridization with three chloroplast-derived probes (ndhA, rbcL and psbA) indicates that 2.25% and 2.7% of the clones in the *Hin*dIII and *Eco*RI libraries, respectively, were derived from the chloroplast genome.

BAC end sequencing

As of March 2000, the CUGI Sequencing Center has sequenced 114 398 BAC ends, or sequence-tagged connectors (STC), from both libraries. 98 853 of these sequences contain > 50 high quality (*phred* 20 or higher) bases. More detailed statistics for the STCs from the *Hin*dIII library are shown in Table 1. The total sequence (52 Mb), and the high quality sequence (25 Mb) obtained from the BAC ends of this library represent 10% and 5% of the rice genome, respectively.

BAC end sequence analysis

A BLASTN search of The Institute for Genomic Research (TIGR) gene indices with the 73 362 STCs of the *Hin*dIII library revealed 11 483 (15.6%) STCs with significant homology (expectation value of 10^{-6} or lower) to at least one expressed sequence tag (EST) in at least one gene index. 9455 (12.8%) of the hits

TABLE 1 Statistics for the BAC end sequences of the Nipponbare *Hin*dIII library

Number of STCs	73 362
Average read length	709
Average HQ bases	339
Total sequenced bases	52 023 659
Total sequenced HQ bases	24 831 896

were to rice ESTs; 10 577 (14.4%) were to either rice or maize ESTs. The 906 STCs (1.2%) that had homologues only in the dicot indices (*Arabidopsis*, tomato or soybean), and not in the monocot gene indices, may constitute the most interesting group of hits, as they may represent genes not previously characterized in monocots.

The STC database of the *Hin*dIII library was examined for transposable element (TE) content by performing a FASTX3 search against a library of 1358 TEs taken from GenBank and a FASTA3 search of the *Hin*dIII library with sequences of known miniature inverted-repeat TEs (MITEs) (cut-off$=10^{-05}$). 9594 BAC end sequences show homology to at least one TE. These include 6027 STCs with homology to retrotransposons, 2746 STCs with homology to MITEs, and 821 STCs with homology to elements that transpose via a DNA intermediate. A more detailed characterization of the TE content of the *Hin*dIII STCs can be found in Mao et al (2000).

The physical map

A physical map of the rice genome was generated by (1) fingerprinting both BAC libraries, (2) assembling the BACs into contigs using *FPC* software, (3) manual editing of the contigs, (4) anchoring the contigs to chromosomes using genetic markers and (5) connecting the contigs with probes derived from contig ends.

BAC fingerprinting

BAC DNA from clones of both libraries was digested with *Hin*dIII and electrophoresed on agarose gels. Gels were stained with SybrGold and captured as tiff images. The *IMAGE* software (Sanger Center) was used to track lanes and identify band sizes. A set of marker DNAs consisting of 20 fragments spanning 32 kb to 1 kb was used to standardize size measurements in all sample lanes. Samples were loaded in duplicate to detect mix-ups during our manual sample preparation and to provide the bandcallers with a choice of lanes to call. All band positions were manually checked, and the copy number of each band was estimated, in order to obtain the highest possible data quality. Comparison of the sequenced BAC OSJNBa0034K24 (Y. Yu et al, unpublished results) with its fingerprint revealed that band size measurements are generally within 1% of the actual fragment size.

Fingerprint analysis

The band sizes determined by the *IMAGE* program are the input for the fingerprint analysis software *FPC* (Soderlund et al 2000). *FPC* uses the Sulston

TABLE 2   Calculated Sulston scores for two hypothetical clones

No. of bands in each clone	Matching bands	Sulston score
1	1	5e-03
5	5	7e-09
10	10	4e-14
10	9	8e-12
10	8	8e-10
20	20	1e-21
20	19	2e-19
20	18	2e-17
20	17	1e-15
20	16	6e-14
20	15	2e-12
20	14	5e-11
20	13	1e-09
20	10	3e-06

The first column shows the number of bands in each clone, the second column indicates the number of matching bands and the third column shows the calculated Sulston score.

score formula (Sulston et al 1988) to run pairwise comparisons of the banding pattern of all clones within the project. Clones that share a large number of bands are more likely to be derived from the same region of the genome than clones that share fewer fragments, and are assigned a lower Sulston score.

Table 2 illustrates the stringency requirements for obtaining a given Sulston score under the conditions we use in our laboratory. Two clones, each containing 10 bands and sharing 9, would not be binned together at a Sulston score of 1e–12. Similarly, two clones containing 20 bands each must share more than 15 bands in order to be binned together. This translates to an overlap of $>90\%$ and $>75\%$, respectively, and illustrates the need for a deep coverage and large insert (many bands per clone) BAC library.

The assembly of fingerprinted clones into contigs takes place in two steps. First, clones that meet a certain stringency requirement are binned together. Next, clones within each bin are ordered based on the extent of overlap with each other. The result is a graphical display of all binned clones in their most likely linear order (contig).

The Sulston score for the autoassembly is determined empirically. The effect of the cut-off on the autoassembly is illustrated in Fig. 1, which includes clones from

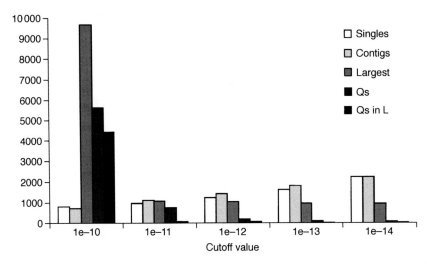

FIG. 1. Effect of the Sulston score cut-off on contig assembly of fingerprinted rice BACs. (33 308 *Hin*dIII clones)

the *Hin*dIII library only. The first two columns show the number of unincorporated singletons and contigs, which increase with increasing stringency (e-10 to e-15). The third column indicates the number of clones in the largest contig of each assembly. At a Sulston score of e-10, almost one third of all clones fall into one contig, indicating that many false overlaps have been established. This is confirmed by the large number of questionable clones (Qs) in the whole project (column 4) and the largest contig (column 5). The Q clones decrease with increasing stringency. The optimal Sulston score for autoassembly is that which provides the fewest falsely connected clones while providing the largest contigs possible.

The *Hin*dIII (36 864 clones) and two-thirds of the *Eco*RI (37 632 clones) library were fingerprinted. Clones with three or fewer bands, or with indication of contamination were not band-called. Over 86% of attempted lanes yielded a usable fingerprint. The rice fingerprint project contains 64 307 clones, which corresponds to > 18 genome equivalents. The exact statistics of this project will continue to change as contigs are edited and revised. Currently, there are 1021 contigs in the project, or about 85 contigs per each of the 12 chromosomes. These contigs span an estimated 436 Mb of sequence, which is close to the DNA content of the rice nuclear genome (430 Mb). In addition, there are 2449 singletons, or unincorporated clones. The average singleton contains only 13.5 *Hin*dIII bands, compared to an average of 28.2 bands for all clones in the project, indicating that most of these clones are not incorporated due to the low number of bands per clone.

Chloroplast contigs

The contig distribution of STCs with homology to chloroplast sequence was examined to assess the quality of the STC and fingerprint databases. The complete STC database containing 127 423 sequences was searched with the 134 525 bp rice chloroplast genome (X15901), using FASTA. BAC ends with at least 61.7% identity over 90 bp were designated 'chloroplast BAC ends' or 'cpSTC'. These parameters were chosen to eliminate false positives resulting from the homology between the nuclear 25S rRNA gene and the large rRNA gene from chloroplast. 3417 cpSTCs from 2320 BACs that had also been fingerprinted were selected in this way.

In the absence of tracking errors and chloroplast-homologous sequences in the nuclear genome, all of these BACs should have been assembled into one contig. In fact, the distribution of these 'chloroplast' BAC ends in contigs is rather complex, as illustrated in Table 3.

13.8% of all clones with chloroplast-homologous BAC ends yielded either no fingerprint (10.9%) or were grouped as singletons (2.9%) in the *FPC* project, indicating that the fingerprint success rate of these clones is comparable to the average for the project (86.2% vs. 86.3%). 79% of all cpSTCs that could be assigned to a contig ended up in one contig, which we have designated the chloroplast contig. The remaining 21% of the clones are distributed in contigs containing from 1–33 chloroplast BAC ends.

This cpSTC distribution is attributable to a number of causes. Mislabelling of clones during either the sequencing or fingerprinting process will result in isolated single or paired cpSTCs, respectively. If we attribute all paired chloroplast STCs that are not in the chloroplast contig to fingerprint tracking error, then this error would be 1.9% (44/2320). It is noteworthy that the 44 pairs of chloroplast BAC ends from contigs with 2–33 chloroplast BAC ends per contig, are distributed as follows: 30 and 10, and 4 and 0 in plates 1–48 and 49–96 of the *Hin*dIII and *Eco*RI library, respectively. This illustrates a gradual error reduction during the course of the project, as the clones were processed in numerical order starting with the *Hin*dIII library. The introduction of automated lane naming in the second half of the *Hin*dIII library was the most significant factor in reducing the error rate.

Unpaired cpSTCs in non-chloroplast contigs (529/3441) can be attributed to a combination of (a) tracking errors during sequencing, (b) chimeric BAC clones formed during BAC library construction or (c) the presence of chloroplast sequences in the nuclear genome. The exact proportion of each component is difficult to assess, but nuclear chloroplast sequences and chimeras undoubtedly account for a share of these clones.

Only 12.3% of the chloroplast BAC end sequences in contigs containing 3–33 cpSTCs are paired (in contrast to 83% pairs in the chloroplast contig).

TABLE 3 Contig distribution of BAC ends with homology to chloroplast sequence (cpSTC)

No. of cpSTCs per contig	No. of BAC end sequences (no. of contigs)	No. of pairs (percentage)
1	117 (117)	NA
2	100 (50)	18 (36)
3	78 (26)	8 (21)
4	36 (9)	5 (28)
5	15 (3)	2 (27)
6	12 (2)	0
7	21 (3)	3 (29)
8	16 (2)	1 (13)
9	36 (4)	3 (17)
10	10 (1)	1 (20)
12	12 (1)	0
15	15 (1)	1 (13)
18	18 (1)	0
20	20 (1)	0
21	21 (1)	0
24	48 (2)	1 (8)
33	66 (2)	1 (3)
2328	2328 (1)	965 (83)
Singletons	99	31 (63)
No fingerprint	373	81 (43)

From left to right are shown the number of cpSTCs per contig, the number of cpSTCs in each category (the number of contigs containing that many chloroplast BAC end sequences), and the number (and percent) of cpSTCs for which the corresponding other BAC end also has chloroplast homology.
NA, not available.

Furthermore, BACs with cpSTCs in contigs containing 3–33 cpSTCs always cluster within each contig, indicating that their fingerprints are very similar. This suggests that these BACs are either chimeric (i.e. contain part nuclear, part chloroplast sequences), or represent regions where chloroplast sequences have integrated into the nuclear genome. In one instance, five clones, each with one cpSTC showing 95% identity over 108 bp to chloroplast DNA, were placed into a contig containing 73 BACs from the nuclear genome. The transfer of organellar genes to the nuclear genome has been documented as an evolutionary force (Martin

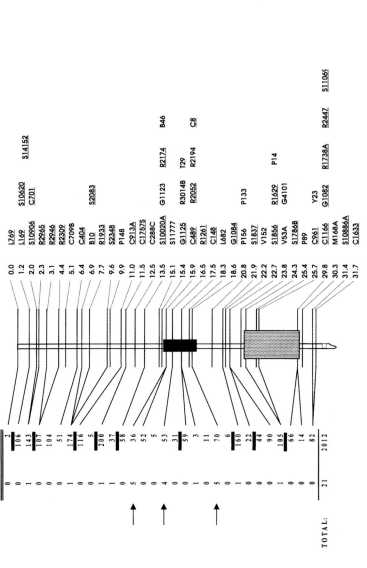

FIG. 2. A map of the top 31.7 cM of chromosome 10 taken from the Japanese Rice Genome website (*http://www.staff.or.jp:80/*) is shown. Markers that were used to probe the BAC libraries and anchor contigs are underlined. The table lists the number of centromere-homologous BAC ends (left column) and total number of BACs (right column) in each of the anchored contigs. Horizontal bars indicate gaps between contigs. The stippled box marks the presumed location of the centromere. The three contigs containing multiple centromere-homologous BAC ends are indicated with arrows, and the chromosomal region they span is marked with a black box. Centromeric sequences used in this search were: AB013613, AB013614, AF058902, AF058903, AF058904, AF058905, AF058906, AF078903, AF091233, Y08025.

& Herrmann 1998, Blanchard & Schmidt 1995), and sequencing of *Arabidopsis* chromosome 2 identified a 270 kb region with 99% identity to the mitochondrial genome (Lin et al 1999).

Anchoring chromosome 10 contigs

The BAC libraries were screened by colony hybridization with 37 validated restriction fragment length polymorphism (RFLP) markers mapped to the short arm and flanking the centromere of rice chromosome 10 by Harushima et al (1998). (Markers were kindly provided by Dr. Takuji Sasaki at the Rice Genome Project, Japan.) All BACs identified in this screen were subjected to Southern hybridization with the respective probe to insure that the BACs contained the mapped restriction fragment. Ambiguous clones were examined by fluorescence *in situ* hybridization (FISH) to verify their chromosomal location. This enabled us to anchor 20 contigs on that chromosome arm. Ten additional contigs were identified using OVERGOs (J. D. McPherson, *http://genome.wustl.edu/gsc/overgo/overgo.html*, Ross et al 1999) designed from end clones of contigs flanking gaps or with probes isolated from the ends of YACs comprising the minimum tile (Kurata et al 1997). Using endprobes, the 30 contigs were further condensed into 9 super-contigs, 7 of which span a total of 17.5cM (segments of 1.1, 1.1, 1.9, 1.9, 2.0, 3.0 and 6.5 cM). The remaining two super-contigs span single or co-segregating markers, thus their size in genetic map distance cannot be determined.

The anchored contigs contain a total of 2012 BACs, which at an average insert size of 125 kb and a 20× genome coverage, can be estimated to span roughly 12.575 Mb. The yeast artificial chromosome (YAC) minimum tile developed by the Rice Genome Project (RGP) for the 0–31.7 cM region of chromosome 10 adds up to 14.25 Mb. Thus our contigs account for 88% of the YAC minimum tile. However, the minimum tile contains several gaps of unknown size, and experience gained in the *Arabidopsis* project illustrates that determining the size of a chromosome using a YAC minimum tile may significantly underestimate the actual DNA content (Lin et al 1999). STC analysis of these contigs (see below) has shown that this 31.7 cM region likely contains the entire centromere and flanking heterochromatin of chromosome 10, which means that some of the remaining gaps may be large and contain repetitive sequences.

Anchoring contigs in silico

We have used the STCs to anchor a significant number of additional contigs throughout the genome *in silico*. Accession numbers for markers from the Japanese RGP and Cornell markers were obtained from the RiceGenes website and used to retrieve the sequences from GenBank. A BLASTN analysis

TABLE 4   Results of a BLASTN search of all RGP and RiceGenes markers against the STC database (cut-off = 10^{-03})

Number of markers with	RGP Japan	RiceGenes — Cornell
accession numbers in RiceGenes	1718	979
accessions in GenBank	1624	548
a match in the BES database	615	188

(expectation value cut-off of 10^{-03}) of these markers against the STC database identified a large number of homologous BAC ends (Table 4). Because of the low cut-off used to identify homologues (sequence homology does not allow distinction between loci), and due to potential tracking errors in both the sequencing and fingerprint databases, caution must be used when incorporating these marker data. However, by combining the marker data with contig information it was possible to anchor a large number of additional contigs genome-wide.

The validity of electronically anchoring BAC contigs was illustrated using a 1.1 Mb p1-derived artificial chromosome (PAC) contig on chromosome 1 that had been sequenced by the Japanese RGP. All markers within that region identified the correct contig as determined by sequence comparison of the PACs with the corresponding STCs.

Throughout the physical mapping project we have concentrated on the Japanese marker set for anchoring contigs. However, the STCs provide an effective means of integrating the Cornell and RGP maps. In addition, we have begun to integrate the rice map with other cereal maps using sequence tags such as the mapped maize markers (MaizeDB: *http://www.agron.missouri.edu/*). At the time of this writing the project contains 897 genetic, OVERGO and YAC end markers, which have been used to anchor 472 contigs covering an estimated 249 Mb to the genetic map.

Why the sidewalk ends . . .

Although fingerprint analysis for the purpose of constructing a physical map has been shown to be very effective, the project is obviously limited by the size of the contigs that can be obtained by this method. The factors affecting contig size are of interest and shall be addressed here.

The most obvious cause for gaps is low representation of some genomic regions in our BAC libraries. We have found several markers that hybridize to far fewer clones than would be expected. For example, marker S1786 identified 19 clones containing the A allele, but only 3 clones with the B allele, indicating that the B allele region is under-represented in the library.

A second reason for gaps between contigs is that some of these regions contain large HindIII fragments, resulting in fewer bands per clone and requiring an unacceptably low cut-off for assembly.

A third obstacle is the presence of duplicated regions. Clones containing duplicated regions are more difficult to assemble properly into contigs. Evidence at the fingerprint, hybridization and sequence level indicates that the rice genome contains significant levels of local duplications. Sequencing of *Arabidopsis* chromosomes 2 and 4 (Lin et al 1999, Mayer et al 1999) also revealed large numbers of local duplications, indicating that this may be a feature of plant genomes.

Localization of Nipponbare centromere 10

Examination of the STC content of anchored contigs yields valuable locus-specific sequence information even before the contigs are sequenced for the genome project. We tried this approach to localize the centromere of chromosome 10. A database of 10 centromeric repeats of rice (Aragon-Alcaide et al 1996, Dong et al 1998, Nonomura & Kurata 1999, Wang et al 2000) was searched with STCs from each of the contigs localized to chromosome 10. The distribution of hits along the chromosome is shown in Fig. 2. Only three contigs contain more than one centromere-homologous STC. The homologies of these STCs are to (a) RCS1, RCB11, RCH2(2) and CCS1-like, (b) RCS2(4), and (c) RCS1(5). All homologies are highly significant (10^{-22} and lower). The area with the highest centromeric repeat density is in the middle of the chromosome arm. It is not yet clear whether this delineates the real centromere (i.e. that it may have been mismapped) or represents a quiescent secondary centromere for this chromosome.

Unexpectedly, only one STC with homology to centromeric elements was found in the contigs located within the mapped location of the centromere. Possibly the centromere is contained in the last gap remaining in that region, and the flanking contigs are not close enough to the centromere to contain centromeric repeat. Alternatively, these results may indicate that the centromere has previously been mismapped.

Conclusion

With the generation of a framework consisting of more than 100 000 STCs and 60 000 fingerprinted BACs, we have assembled the tools required for efficient sequencing of the rice genome. Integration of these tools with those developed by the Japanese RGP and others will facilitate rapid genome sequencing. Valuable information can be gained by examining the STC content of anchored contigs even before genome sequencing is complete.

Acknowledgements

Funding for construction of the BAC libraries was provided by The Rockefeller Foundation. The BAC end sequencing and fingerprinting was funded by Novartis. Anchoring of contigs to the genetic map by hybridization was funded by Novartis and the USDA/NSF/DOE Rice Genome Sequencing Program (NSF DBI-9982594 and USDA 99-35317-8505). We thank T. Sasaki of the Japanese RGP for kindly providing the Nipponbare seed for BAC library construction and the chromosome 10 markers.

References

Aragon-Alcaide L, Miller T, Schwarzacher T, Reader S, Moore G 1996 A cereal centromeric sequence. Chromosoma 105:261–268

Blanchard JL, Schmidt GW 1995 Pervasive migration of organellar DNA to the nucleus in plants. J Mol Evol 41:397–406

Dong F, Miller JT, Jackson SA, Wang GL, Ronald PC, Jiang J 1998 Rice (*Oryza sativa*) centromeric regions consist of complex DNA. Proc Natl Acad Sci USA 95:8135–8140

Harushima Y, Yano M, Shomura A et al 1998 A high-density rice genetic linkage map with 2275 markers using a single F2 population. Genetics 148:479–494

Kurata N, Umehara Y, Tanoue H, Sasaki T 1997 Physical mapping of the rice genome with YAC clones. Plant Mol Biol 35:101–113

Lin X, Kaul S, Rounsley S et al 1999 Sequence and analysis of chromosome 2 of the plant *Arabidopsis thaliana*. Nature 402:761–768

Mahairas GG, Wallace JC, Smith K et al 1999 Sequence-tagged connectors: a sequence approach to mapping and scanning the human genome. Proc Natl Acad Sci USA 96:9739–9744

Mao L, Wood TC, Yu Y et al 2000 Rice transposable elements: a survey of 73 000 sequence-tagged-connectors. Genome Res 10:982–990

Martin W, Herrmann RG 1998 Gene transfer from organelles to the nucleus: how much, what happens, and why? Plant Physiol 118:9–17

Mayer K, Schuller C, Wambutt R et al 1999 Sequence and analysis of chromosome 4 of the plant *Arabidopsis thaliana*. Nature 402:769–777

Nonomura KI, Kurata N 1999 Organization of the 1.9-kb repeat unit RCE1 in the centromeric region of rice chromosomes. Mol Gen Genet 261:1–10

Ross MT, LaBrie S, McPherson J, Stanton VP Jr 1999 Screening large-insert libraries by hybridization. In: Boyl A (ed) Current protocols in human genetics. Wiley, New York, p 5.6.1–5.6.52

Soderlund C, Humphray S, Dunham A, French L 2000 Contigs built with fingerprints, markers and FPC v4.7. Genome Res 10:1772–1787

Sulston J, Mallet F, Staden R, Durbin R, Horsnell T, Coulson A 1988 Software for genome mapping by fingerprinting techniques. CABIOS 5:125–132

Wang S, Wang J, Jiang J, Zhang Q 2000 Mapping of centromeric regions on the molecular linkage map of rice (*Oryza sativa* L.) using centromere-associated sequences. Mol Gen Genet 263:165–172

DISCUSSION

Gale: Quite often we hear that unlike YACs, BACs are not chimeric. What percentages of your BACs are chimeric?

Presting: We don't have precise data on how many BACs are chimeric. Our results indicate that some BACs may be chimeric, but we cannot distinguish with certainty between chimeric BACs and chloroplast-derived nuclear sequences.

Ku: Do you know the functions of these chloroplast sequences in the nuclear genome?

Presting: No. I'm not sure either what the mitochondrial genome is doing in the *Arabidopsis* centromere; we just know that it is there. It may have been trapped amongst the chromosomes during cell division, become enveloped by the nuclear envelope and subsequently integrated into the chromosome some time during evolution.

Ku: Was the whole gene sequenced, or just a fragment?

Presting: We only have BAC end sequence data; everything I said relates to sequences of 500 bp and under.

Mazur: What is the evidence that the chloroplast DNA is not a chimeric clone?

Presting: That is a good question. We have one set of contigs that contains all the chloroplast clones, and then we have a whole bunch of other contigs, about 50, which seem to contain both chloroplast sequences and nuclear sequences.

Mazur: Since chloroplast sequences are a frequent part of the DNA population, you might expect that they would be a frequent source of chimeras.

Presting: That is absolutely right. What I have shown here is that we have identified BAC clones that are likely to be chimeric. To prove that they are chimeric, we would first need to verify that they are not the result of tracking errors. I suppose we could then genetically or physically map both ends of these BACs. If the BAC is chimeric, the ends should not map near each other. Another way to determine the amount of chloroplast sequence in the nuclear genome is by FISH. We therefore intend to take some of the BACs that we know contain 100% chloroplast DNA and hybridize them in FISH experiments to see if we can detect them in the nuclear genome.

Parker: You mentioned the gene duplication analysis. What is the extent of gene duplication in the rice genome?

Presting: The most extensive analysis that we have done is on the short arm of chromosome 10, which is fairly heterochromatic. Often the reason that BAC contigs don't merge appears to be because of aggregation of clones at the ends of the contigs, presumably because they contain duplicated regions. I'm not sure how common gene duplications are in rice, but in *Arabidopsis* they are fairly extensive. The group at TIGR discovered around 250 duplications on chromosome 2. Some were ancient and some were very recent. I suspect we will find something similar in rice, which will make sequencing a little more difficult.

G.-L. Wang: You have used a fingerprinting method to construct the contigs in the rice genome. How many enzymes did you use to digest the BACs for those fingerprints?

Presting: That is an important point. If you produce your fingerprints with the same enzyme that you use for cloning, you can remove the vector band. Thus all the bands you include in the analysis are genomic bands. We were torn between fingerprinting the *Hin*dIII library with *Hin*dIII and *Eco*RI library with *Eco*RI and then connecting the two datasets with marker hybridization data. In the end we just decided to use *Hin*dIII to cut the *Eco*RI library as well.

G.-L. Wang: Can you use *Bam*HI to digest the BACs for fingerprinting, in order to separate the genomic insert from the vector?

Presting: I believe that pBeloBAC11 and pBACIndigo each contain only one *Bam*HI site, so you would not be able to liberate the insert with *Bam*HI.

G.-L. Wang: Since you have a lot of BAC ends sequenced, can you use that as a probe to map to the linkage map as a marker? That way you could fill a lot of the gaps in the rice genome.

Presting: Yes. In fact, the RGP has been very helpful in placing some BAC end sequences onto the genetic map. We also make extensive use of a technique called OVERGO hybridization in which we design short 40 bp oligos from the BAC end sequence and then use these for hybridizations, both against BAC filters and for Southerns.

Matsumoto: I have a question about the 10% contamination of the retrotransposon sequences in the BAC end sequences. Do you think that there is some relationship between the *Hin*dIII site and retrotransposon existence?

Presting: All the sequence information we have discussed relates to the fact that there has to be a *Hin*dIII site present. It will be interesting to see what percentages of BAC ends in the *Eco*RI library contain retrotransposons or transposable elements.

Matsumoto: In our genomic sequence in 47 PACs, we find that the frequency of retrotransposon sequences is very high. There is almost one retrotransposon in each PAC sequence. Perhaps this high percentage from your BAC end sequence is not a surprise.

Li: Where you have large coverage of the physical map of the genome, have you found any discrepancies in terms of the gene and marker orders between your physical map and the published linkage map? In the physical map where you have the contig and the marker, can you link them? On the basis of the physical map, is there any inversion or other discrepancies with the linkage map?

Presting: I don't think we have enough data at this point to know. What we do know is that some data do not make sense. We are looking at the possibility of some local inversions in the map. But the vast majority of the marker data are exactly the way they should be. The only markers that we have had to rearrange so far are markers that co-segregate, where we have not been able to determine marker order genetically.

Gale: You indicated that there just might be a difference between the indica and the japonica maps, in relation to the location of the centromere. It is pretty worrying when we find that the positions of the centromeres are well matched between rice, wheat and maize! Perhaps your observation could just reflect an inversion.

Presting: I think it could just be a localized inversion, and this is why I was so careful in what I said. So far we have only indirect evidence from the BAC end sequences. We don't know for sure that the centromeres are in different locations. It is just one explanation of the discrepancy.

Khush: The original centromere was mapped on the indica, and then the position was confirmed on the japonica using the japonica markers from the rice genome mapping project. I find it difficult to believe that there will be differences.

Bennett: Do you expect that you will be able to complete the physical map using the same strategy and the same two libraries, or will you need to introduce additional approaches?

Presting: The key to finishing the physical map will be to integrate as many resources as are available, particularly in Japan where a lot of PAC clones have been anchored throughout the genome. If, for example, we can fingerprint those PAC clones and assign them to a contig, we will have automatically anchored this contig to where the PAC clone is anchored. Integration of data is going to be crucial.

Rice genomics: current status of genome sequencing

Takashi Matsumoto, Jianzhong Wu*, Tomoya Baba*, Yuichi Katayose, Kimiko Yamamoto*, Katsumi Sakata, Masahiro Yano and Takuji Sasaki

*Rice Genome Research Program, National Institute of Agrobiological Resources, 2-1-2 Kannondai, Tsukuba, Ibaraki 305-8602, and *STAFF Institute, Ippaizuka, Kamiyokoba, Tsukuba, Ibaraki 305-0854, Japan*

Abstract. Since its establishment in 1991, the Rice Genome Research Program (RGP) has produced some basic tools for rice genome analysis, including a cDNA catalogue, a genetic linkage map and a yeast artificial chromosome (YAC)-based physical map. For the further development of rice genomics, RGP launched in 1998 an international collaborative project on rice genome sequencing. A P1-derived artificial chromosome (PAC)-based, sequence-ready physical map has been constructed using the PCR markers from cDNA sequences (expressed sequence tag [EST] markers). Selected PAC clones with 100–150 kb inserts from chromosomes 1 and 6 have been subjected to shotgun sequencing. The assembled genomic sequences, after predicting the gene-coding region, have been published both through a public database and through our website. As of January 2000, 1.9 Mb from 13 PAC clones were published. Future prospects for understanding rice genomic information at the nucleotide level are discussed.

2001 Rice biotechnology: improving yield, stress, tolerance and grain quality. Wiley, Chichester (Novartis Foundation Symposium 236) p 28–41

Rice (*Oryza sativa* L.) is one of the world's major crops; it is the staple food for about half the population of the world. The rate of yield increase in rice production has been declining for the past decade (Conway & Toenniessen 1999) while the world population will still be increasing in the 21st century, reaching an estimated 9–10 billion persons by 2050 (United Nations 1998). To achieve a substantial improvement in grain yield in a limited period, a 'second green revolution' based on advanced plant biotechnology and plant genomics is needed.

Rice genomics is important not only for rice biology and biotechnology but also for other cereals because rice has the smallest genome size of the major cereals (Arumuganathan & Earle 1991), and has high synteny with other cereal crops (Ahn & Tanksley 1993, Kurata et al 1994, Moore et al 1995). Recently, the genomic DNA sequences from two chromosomes of a dicot plant, *Arabidopsis thaliana*, were published (Lin et al 1999, Mayer et al 1999). They are expected to contribute to the understanding of basic biological mechanisms in higher plants. However, it is still not certain whether all phenotypic traits of cereal crops, which are important targets for breeding, can be elucidated by *Arabidopsis* gene functions. An attempt to find colinearity between *Arabidopsis* and rice genomes using rice expressed sequence tags (ESTs) as probes did not succeed (Devos et al 1999). This indicates that rice genomics is also important for the comparative analysis of monocots and dicots, and thus for basic research in plant biology.

The Rice Genome Research Program (RGP) was initiated in 1991 with the objective of understanding the whole rice genome. Since then it has set up fundamental resources and databases for rice genome analysis, such as a cDNA catalogue of about 40 000 expressed genes, a genetic linkage map with 2275 DNA markers (Harushima et al 1998), and a yeast artificial chromosome (YAC)-based physical map covering 63% of the genome (Saji et al 2001). In 1998, the project was reorganized as a new program with three main goals: (1) complete genome sequencing, (2) elucidation of gene functions, and (3) application of genome information to breeding. This paper describes the strategy, system construction and performance of the project, and the present status of genome sequencing of rice chromosomes 1 and 6.

Strategy for genome sequencing

Using technological improvements in the sequencing apparatus, such as high-throughput PCR machines, capillary DNA sequencers, and computer software that can assemble thousands of sequence fragments, we can now reconstitute the whole sequence of a 100–150 kb long insert by the 'shotgun' sequencing method. The shotgun method has the advantage that it takes less time overall than other methods to make a long sequence contig; also, because of the high redundancy in the method, the resulting sequence is more accurate than that from one-pass sequencing.

On a genome-wide scale however, there is controversy concerning the effectiveness of whole-genome shotgun sequencing (Weber & Myers 1997, Green 1997, Little 1999). This approach, which is based on sequencing and assembly of unselected clones, has been used for microbes (for example, Fleischmann et al 1995), and has now been applied to higher organisms, such as the fruitfly genome and the human genome. As a test case, the whole-genome shotgun sequencing of the fruitfly *Drosophila melanogaster* has been recently completed and over 45 Mb of this sequence was deposited to public database (announced from the website of the Berkley *Drosophila* Genome Project, BDGP: *http://www.fruitfly.org/*). This approach has produced many sequence contigs whose locations were unknown. In the *Drosophila* genome, such contigs could be identified by using the bacterial artificial chromosome (BAC)-based physical map produced by BDGP. Therefore, we believe this whole-genome shotgun approach to the genomes of higher organisms is valid only when it is combined with a clone-by-clone approach. Moreover, whole-genome sequence assembly may result in global misassemblage that causes confusion. Thus, we chose a stepwise, clone-by-clone sequencing approach for the rice genome.

In this approach, the rice genome is divided into several thousand relatively large chromosomal segments, which are then cloned by P1-derived artificial chromosome vectors (PAC library). These PAC clones are aligned to their chromosomal positions based on the EST markers. The position-identified PAC clones are selected and subjected to shotgun sequencing. The sequence data are assembled to form continuous sequences. This local shotgun approach has the advantage of easily finding misassembled contigs, producing accurate sequence data.

EST marker establishment based on the YAC physical map

A sequence-ready physical map of BAC, PAC, etc., is indispensable to our strategy. To make a comprehensive physical map, DNA markers that are mapped at a high density are required. Although the RGP has constructed a YAC-based physical map using 2275 DNA markers (restriction fragment length polymorphism [RFLP] markers and PCR markers), this number of markers is far from sufficient to construct a sequence-ready physical map, because the average size (100–150 kb) of the genomic inserts in the BAC or PAC is smaller than that of the average YAC (350 kb in the case of the RGP YAC library). In order to generate more markers in a relatively short time, we have begun comprehensive mapping of cDNA clones as PCR markers using gene-specific primers from 3′-untranslated regions (UTRs) and the YAC-based physical map. About 7000 YAC clones (Umehara et al 1995) have been pooled and PCR screening with primers on the pooled clones has been carried out. More than 90% of the primers from 3′-UTRs have amplified a single band, showing gene-specific amplification. At present, about 4500 of these physically mapped ESTs and 970 sequence-tagged site (STS) markers (derived from RFLP markers) have been established throughout the genome for anchoring PAC clones.

Construction of the PAC genomic library and PAC physical contigs

The PAC is a kind of BAC system. This vector can retain large (more than 100 kb) DNA fragments in *Escherichia coli* in a stable manner (Ioannou et al 1994). The major advantage of this system is the positive selection of recombinant clones by using a suicide gene, *sac*B. The RGP has constructed a PAC library from young leaves of *Oryza sativa* L.cv. Nipponbare (Baba et al 2000). Genomic DNA was partially digested with *Sau*3AI and cloned into the *Bam*HI site of the PAC vector, pCYPAC2. Figure 1 shows the insert size determination of recombinant PAC clones analysed by pulsed-field gel electrophoresis. The total number of clones was 71 000 and the average insert size was 112 kb, providing 16-fold coverage of the rice genome. Using PCR primers from EST (and STS) markers, this library underwent high-speed PCR screening in 96-sample base. In all, 8769 PAC clones were screened with 2075 EST markers and 252 STS markers, and formed 798 PAC contigs over the whole genome. These contigs covered about 30% of the rice genome, including 17 Mb of chromosome 1 and 13 Mb of chromosome 6. Clone overlaps within a contig in the short arm of chromosome 1 were confirmed by restriction enzyme digestion fingerprinting (Coulson et al 1986) before the clones were subjected to shotgun sequencing.

Shotgun DNA sequencing and sequence assembly

Selected PACs were subjected to shotgun sequencing. From 200 ml bacterial culture, 5 to 20 μg PAC DNA were purified by the alkaline lysis method followed by ultracentrifugation. Purified PAC DNA was sheared by sonication and separated by agarose-gel electrophoresis. The 2 kb and 5 kb fragments were recovered from the agarose gel, blunt-ended, and ligated into the *Sma*I site of the pUC18 vector. The ligated DNA was used for transformation of the DH10B strain of *E. coli*. Two kinds of subclone libraries (2 kb insert library and 5 kb insert library) were constructed. For one PAC, about 3000 colonies — 2000 from the 2 kb library and 1000 from the 5 kb library — were collected by a robotic picking machine and cultured on a 1 ml scale. Plasmid DNAs were treated with an automated plasmid purification system using the alkaline lysis method (Birnboim & Doty 1979) on a 96-sample base. About 500 bases from both ends of the subclone inserts were sequenced by the Big Dye Terminator method (PE Biosystems) using a capillary sequencer (PE 3700).

Sequences from 1000 2 kb plasmid templates and 1000 5 kb plasmid templates were used for making the major contigs. Additional templates were sequenced if the contigs analysed had more than five gaps. Sequences from 5 kb templates that bridged two unconnected contigs were also used for gap closure. At the 'finishing' stage, gaps were filled by full sequencing of the bridge clones using the primer-walking, shattered-library (McMurry et al 1998) or nested-deletion methods (Hattori et al 1997). Low-quality data, as judged by the Tracetuner (Paracel Inc.)-phrap software (Ewing & Green 1998), were improved to an international standard (i.e. error rate less than one per 10 000 bases). The overall procedure is shown in Fig. 2.

We have noticed that the rice genome has many dinucleotide repeat sequences such as TA, GA and TC and clusters of G or C. The number of these repeats and clusters in a PAC varies from several to 40. These have been the difficult regions to

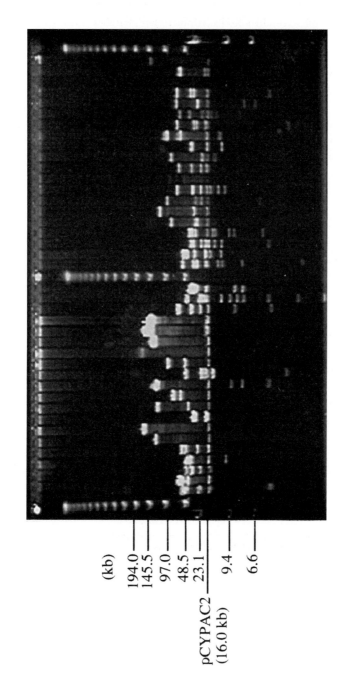

(kb)

194.0
145.5
97.0
48.5
23.1

pCYPAC2
(16.0 kb)

9.4

6.6

FIG. 1. Size determination of PAC clone inserts by pulsed-field gel electrophoresis. PAC DNAs are purified from each PAC clone and digested with NotI. The digested DNA is subjected to electrophoresis along with molecular size markers (lambda phage DNA digested with HindIII and lambda ladders). The insert size is calculated as the sum of the length of the fragments, except for the PAC vector (indicated as the arrow in the figure).

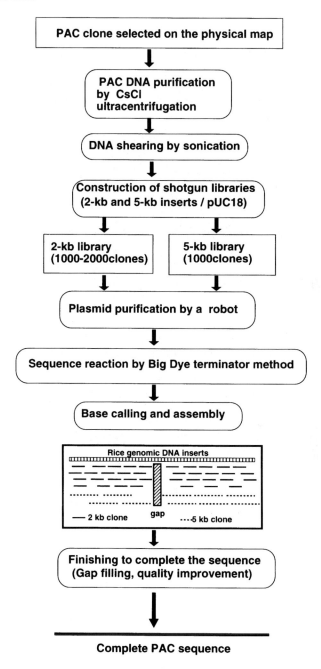

FIG. 2.   Flowchart of the genome sequencing procedure at RGP.

sequence. Sequence reactions would often stop at or near these sequences. This may be caused by the secondary structure formed by such sequences. Sometimes different subclones have different sizes of repeats in the same region. This apparent deletion/insertion may be caused by slippage of DNA polymerase. Most of these difficult sequences have been resolved by using several sequencing chemistries other than the Big Dye Terminator, such as Big Dye Primer, dRhodamine Terminator, dGTP Terminator (PE Biosystems), and Thermo Sequenase Dye Terminator (Amersham). Where some short sequence gaps existed or sequence differences were observed between subclones, sequence reactions using PAC DNAs as templates were effective. These repeat or GC-biased regions were locally concentrated, and the total GC content of a PAC genomic sequence was 40–45% (the overall GC content of two chromosomes of *Arabidopsis* is 36%).

Annotation and data publication

The PAC genome sequence is subjected to an annotation step, in which protein-coding regions are predicted in two ways. First, the genomic sequence is searched for similarity in the databases (the rice cDNA database at RGP and a non-redundant protein database at the MAFF DNA Bank) with the BLAST algorithms (Altschul et al 1994)). Next, exon/intron junctions are predicted by use of prediction software (Splice Predictor [Brendel & Kleffe 1998] and Genscan [Burge & Karlin 1997]). We apply to rice genomic data both the *Arabidopsis* and maize rules for the identification of the coding frame.

These data are integrated, edited and plotted as annotations. Finally the annotated sequences are made public through DDBJ deposition and our website through the INE (INtegrated rice genome Exploler; Sakata et al 2000) software, which is a relational genome information database (*http:// rgp.dna.affrc.go.jp/giot/INE.html*). Figure 3 shows a typical example of the output of our web-based publication.

Current status

As of January 12th, 2000, 1 935 409 bases from 13 PAC clones (seven from chromosome 1 and six from chromosome 6) have been published. The introduction of capillary sequencers has enhanced our productivity. We have constructed a system that processes 8–10 PACs per month, from PAC clone selection to sequence publication. The total predicted number of genes from these sequences is 355, including 13 transposon-encoded sequences. Therefore, the overall gene density of these regions is one gene per 5.5 kb. This gene density is almost the same order as that of *Arabidopsis* (4.4 kb and 4.6 kb per gene for chromosomes 2 and 4, respectively). From gene density, the total number of

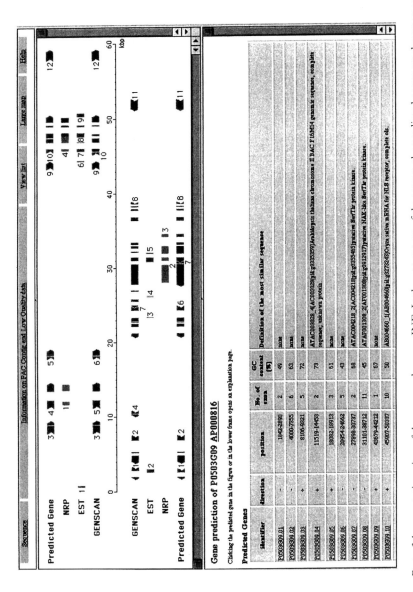

FIG. 3. Part of the annotation view of the genome browser INE. In the upper part of the screen, the predicted genes are shown as arrows in the predicted positions. In the lower part, tables that provide detailed information on the predicted genes are shown.

genes in *Arabidopsis* is roughly estimated to be 31 000 (genome size estimated as 140 Mb). Considering that the genome of rice is about three times larger than that of *Arabidopsis*, the gene density found on the PAC sequences might be overestimated. Since 1.9 Mb is only 0.4% of the whole rice genome, we are not sure whether this gene density applies to the whole genome. One could speculate that because we selected PACs with DNA markers that were mostly derived from cDNAs, gene-rich regions may have been sequenced predominantly. Further results of genomic sequencing will answer this question.

International collaboration

Rice genome sequencing at the RGP is part of a multinational research project (International Rice Genome Sequencing Project) that aims to complete the sequencing of the rice genome in 10 years. For this purpose, resources are shared with the members of the research consortium; the Nipponbare rice variety is used as the sole plant resource for the project. The RGP PAC library and the BAC library constructed at Clemson University, USA, have been endowed for genome sequencing. There is a 'chromosome sharing' in which each country of the consortium has responsibility for sequencing at least one megabase per year of the shared chromosomes. The RGP proposes, as its contribution to this project, the sequencing of chromosomes 1 and 6. More information on the international project is available in the electronic newsletter, Oryza (*http://rgp.dna.affrc.go.jp/rgp/News/Newsletter.html*).

Future prospects

30% of the total genome has been covered by PAC clones, but still many regions remain as gaps. These gaps on the physical map might be filled by the walking strategy based on the PAC end-sequences. One possible reason for these gaps is that some of the *Sau*3AI recognition sequences might be methylated and cannot be digested. To complement our PAC library we have constructed a new BAC library utilizing *Mbo*I as an enzyme for partial digestion; *Mbo*I works at the methylated site. The speed of sequence production depends more and more on investment in new technology. New cutting-edge technology will revolutionize DNA sequencing and minimize the time required to complete sequencing of the rice genome.

   In the future, all sequence information could be deposited in one common database for public use. A centre for genome information analysis, which opened in 1999 at the National Institute of Agrobiological Resources, Japan, aims to function as the central database of rice genomic information. At this centre, genomic sequences from all the consortium members could be re-annotated. The ongoing prediction program is not necessarily valid for rice gene prediction because it was developed for other plants. Therefore, computer software for rice gene prediction has been developed based on our coding sequence (cDNAs) data (K. Sakata, unpublished data). Information on the whole rice genome will help isolate genes for biologically and agronomically important traits and help making DNA markers for breeding at the desired place on the genome. Rice genome sequences, along with those of other important cereals, will obviously contribute to overcoming food deficiency in the 21st century.

Acknowledgements

We thank our many colleagues in the RGP for collaborating in this study. The work was supported by the Ministry of Agriculture, Forestry and Fisheries of Japan.

References

Ahn S, Tanksley SD 1993 Comparative linkage maps of the rice and maize genomes. Proc Natl Acad Sci USA 90:7980–7984

Altschul S, Boguski MS, Gish W, Wootton JC 1994 Issues in searching molecular sequence databases. Nat Genet 6:119–129

Arumuganathan K, Earle ED 1991 Nuclear DNA content of some important plant species. Plant Mol Biol Reporter 9:208–218

Baba T, Katagiri S, Tanoue H et al 2000 Construction and characterization of rice genomic libraries: PAC library of *Japonica* variety, Nipponbare and BAC library of *Indica* variety, Kasalath. Bull Publ Natl Inst Agrobiol Resour 14:41–49

Birnboim HC, Doty J 1979 A rapid alkaline extraction procedure for screening recombinant plasmid DNA. Nucleic Acids Res 7:1513–1523

Brendel V, Kleffe J 1998 Prediction of locally optimal splice sites in plant pre-mRNA with applications to gene identification in *Arabidopsis thaliana* genomic DNA. Nucleic Acids Res 26:4748–4757

Burge C, Karlin S 1997 Prediction of complete gene structures in human genomic DNA. J Mol Biol 268:78–94

Conway G, Toenniessen G 1999 Feeding the world in the twenty-first century. Nature (suppl) 402:c55–c58

Coulson AS, Sulston J, Brenner S, Kam J 1986 Toward a physical map of the genome of the nematode *C. elegans*. Proc Natl Acad Sci USA 83:7821–7825

Devos KM, Beales J, Nagamura Y, Sasaki T 1999 *Arabidopsis*–rice: will colinearity allow gene prediction across the eudicot–monocot divide? Genome Res 9:825–829

Ewing B, Green P 1998 Base-calling of automated sequencer traces using phred. II. Error probabilities. Genome Res 8:186–194

Fleischmann RD, Adams MD, White O et al 1995 Whole-genome random shogun sequencing and assembly of *Haemophilus influenzae* Rd. Science 269:496–512

Green P 1997 Against a whole-genome shotgun. Genome Res 7:410–417

Harushima Y, Yano M, Shomura A et al 1998 A high-density rice genetic linkage map with 2275 markers using a single F_2 population. Genetics 148:479–494

Hattori M, Tsukahara F, Furuhara Y et al 1997 A novel method for making nested deletions and its applications for sequencing of a 300 kb region of human APP locus. Nucleic Acids Res 25:1802–1808

Ioannou PA, Amemiya CT, Garnes J et al 1994 A new bacteriophage P1-derived vector for the propagation of large human DNA fragments. Nat Genet 6:84–89

Kurata N, Moore G, Nagamura Y et al 1994 Conservation of genome structure between rice and wheat. Bio-Technology 12:276–278

Lin X, Kaul S, Rounsley S et al 1999 Sequence and analysis of chromosome 2 of the plant *Arabidopsis thaliana*. Nature 402:761–768

Little P 1999 The book of genes. Nature 402:467–468

Mayer K, Schüller C, Wambutt R et al 1999 Sequence and analysis of chromosome 4 of the plant *Arabidopsis thaliana*. Nature 402:769–777

McMurray AA, Sulston JE, Quail MA 1998 Short-insert libraries as a method of problem solving in genome sequencing. Genome Res 8:562–566

Moore G, Devos KM, Wang Z, Gale MD 1995 Cereal genome evolution. Grasses, line up and form a circle. Curr Biol 5:737–739

Saji S, Umehara Y, Antonio BA et al 2001 A physical map with yeast artificial chromosome (YAC) clones covering 63% of the 12 rice chromosomes. Genome 44:32–37

Sakata K, Antonio BA, Mukai Y et al 2000 INE: a rice genome database with an integrated map view. Nucleic Acids Res 28:97–101

Umehara Y, Inagaki A, Tanoue H et al 1995 Construction and characterization of a rice YAC library for physical mapping. Mol Breed 1:79–85

United Nations 1998 World population estimates and projections. 1998 briefing packet. United Nations Population Division, New York

Weber JL, Myers EW 1997 Human whole-genome shotgun sequencing. Genome Res 7:401–409

DISCUSSION

Leung: Do you have any ideas about a fast-track to the application of EST data in terms of associating the EST information with biological variation? On the basis of EST sequences, is there enough information in designing STS primers to pick up variability in the germplasm? Currently we are very much relying on microsatellites in terms of looking at differences between varieties. Is it possible to use existing EST sequences to look at variations in germplasm stock? It would be very powerful to align the physical EST map with allelic variation that can link to certain biological variations.

Matsumoto: I have no idea. Our EST map is based on the 3′-UTR region, so it is very specific to genes. Almost all the genes could give single band amplification.

Presting: I don't have any specific comments regarding ESTs, but I do have a more general point about data integration. I think that allelic variation is an excellent theme to incorporate into a database and make visible to everybody. This would involve integration of the physical and the sequencing map. We have to think about how this should be done. Accessibility of the data and open communication are key issues.

Dong: Do you have a mechanism to update your annotated sequence? In *Arabidopsis* we notice that annotation can often be wrong, so it is actually misleading.

Matsumoto: We have some rules for submitting genomic sequences with annotations. We are going to re-annotate our genomic sequence by using updated EST sequences. One of the purposes of the centralized database is to reannotate our sequence and hopefully the other regions with a unified standard.

Okita: It has been estimated that it will take seven years to sequence the rice genome. Given the technologies available to the private sector, what would be a more realistic estimate for the complete genome sequence of rice?

Matsumoto: You saw our progression in sequencing. The curves are bimodal. The turning point was the introduction of capillary sequencers. The introduction of new technology is very important in reducing the time taken to sequence the rice genome, and the genome will almost definitely be sequenced before 2007. However, we don't have any information about what is going on in Celera and other such companies.

Mazur: You commented accurately on the difficulty of the whole genome shotgun approach, given all the repeats. Is anyone doing a shotgun strategy for rice?

Matsumoto: I attended a meeting run by The Institute for Genomic Research (TIGR) in September 1998, and Eugene W. Myers of Celera showed the Celera assembler, using Compaq computers. They distinguished every single repeat. There might be a one base pair difference per 200 bases in each repeat, and the Celera assembler can distinguish that difference, in order to locate each repeat in the appropriate position. However, this was in *Drosophila*, not rice. We don't have access to the same computing power that Celera does.

Mazur: So they haven't tried assembling rice.

Matsumoto: That is correct.

Nevill: A comment from the industry perspective: the interest in sequencing the rice genome is not just for the sake of rice, but also because of the possibility of extrapolating from this into other cereals. On this basis, therefore, I think it is worth putting a lot more effort into getting the sequence earlier, perhaps not with the highest level of precision.

Goff: The public projects are aiming at 99.99% accuracy, which costs about US$0.60 per base. The rice genome would therefore cost about US$225 million. Is it worth this much to any company? There may be approaches to gene discovery that are less costly, the goal of which would not be 99.99% accuracy. At US$0.60 a base, maize would cost approximately US$1.5 billion, and wheat would cost five times that amount.

Mazur: I agree. I don't think our costs are that high, but we don't intend to do a complete sequence of rice, maize or any other crop.

G.-L. Wang: Dr Matsumoto, let's say you have a 500 kb sequence, and from your prediction you may have 100 genes in this region. How many ESTs can you find in your EST database?

Matsumoto: About 50%. Our ESTs did not hit the rest of them.

G.-L. Wang: Do you have any project or initiative to do the transcription map to find the missing 50% of genes?

Matsumoto: Not at present. Our colleagues are planning to sequence more ESTs. This is the full-length cDNA project of rice. They intend to make more ESTs to supplement our EST database.

Gale: What percentages of your ESTs give unambiguous single hits? Presumably a lot of them are duplicated.

Matsumoto: About 60% of the ESTs that we have tried using are mapped onto specific regions of the chromosomes. 90% of ESTs gave a single-band amplification, but when we mapped the amplified band, some 30% came to the duplicate position. The final success rate of the mapping is 60%.

Gale: Is that your present best guess of how much of the rice genome is duplicated? Gernot Presting was earlier asked the same question, as to what extent the rice genome is duplicated. Presumably it is going to be worse in rice than in *Arabidopsis.*

Matsumoto: I am not sure. In our case we don't use the duplicated ESTs. If we find evidence of duplication, we don't use these ESTs for PAC screening.

Goff: Genomics companies spend 10–20% of their effort on bioinformatics. What is the public perception of how much money is going into bioinformatics, and is this rate-limiting for analysis of the data generated?

Matsumoto: I cannot tell you how much money is going into it, but bioinformatics is certainly very important. We have several projects ongoing in parallel, each of which generates an enormous amount of information. No one knows the whole figure. We don't have great facilities for informatics in Japan.

Nevill: You briefly mentioned functional genomics. How much effort do you put in at the moment to looking at the transcript or at the proteomics level?

Matsumoto: We have just begun doing this. We have transposon mutant panels, which have identified certain genes. Some of these genes were in our genomics sequence, thus we have combined the results of the genomic sequence and the results of mutant panels. Because the sequenced region is relatively small at the moment, only about 2% of the total genome, the probability of finding the genome region is small. We have just begun a microarray analysis using our ESTs.

Okita: Could you briefly describe what effort is being put into proteomics in Japan?

Matsumoto: I'm not a proteomics person, so I'm probably not the best person to ask. The proteomics programme has just begun, using 2D gels with separated proteins in various conditions. Part of the project involves the 3D structure determination of important gene products.

Gale: Earlier on we saw the advance of the rice sequencing project. The contributions being made by different countries and the different chromosome regions they are going to sequence indicate a pretty wide international effort. However, at the moment, as far as I am aware there is only a little bit of Korean money, a small amount of US money and the Japanese effort. What do you think the prognosis is? Are we really going to get other countries to come in?

Matsumoto: Other countries have been assigned the task of sequencing various chromosomes. In the last year they started to sequence several PACs and BACs. We have a target of sequencing one megabase each year for each country to stay in the sequencing club. There are now 10 international sequencing projects. The USA had only one chromosome last year, but now they have added chromosomes 3 and 11. The 12 chromosomes could be sequenced by 2007. The assignment of the chromosomes may be changed. For example, India was originally assigned chromosome 8, but has moved to chromosome 11, in collaboration with the USA. The UK was assigned chromosome 2.

Gale: I saw the UK flag up there, which probably should have been a European flag, but we have failed to get the money this time round from the EU, and I rather doubt that we will be successful within 'Framework V'. What is the situation with the other countries on that list?

Matsumoto: Taiwan, India and Thailand are funded by their governments. We are hopeful that the rice genome sequence will be seen as important by funding agencies.

Khush: France was supposed to take one chromosome. Have they started the work?

Matsumoto: Yes, this is chromosome 12.

Presting: In the USA we have been given US$12 million by the USDA, NSF and DOE. This is what funds the public effort, which initially was chromosome 10, and has now expanded to chromosome 3. There are also privately funded efforts at Rutgers and the University of Wisconsin. The hope must be that people will get drawn into sequencing more as the resources improve. Our initial goal was just to provide a physical map for ourselves, but now it seems that we have the capability to expand it genome-wide. Not in the same detail as for our chromosomes (10 and 3), but well enough to provide a useful tool for others who want to start sequencing other chromosomes. I have received a lot of hybridization data, especially from our British colleagues. Ian Bancroft's hybridization data have allowed us to anchor roughly 30 contigs on chromosome 2. We can only hope that this collaborative attitude among the different projects will prevail, as it will advance the sequencing effort as a whole. By providing contigs for the other chromosomes, we might make the project more attractive to the funding agencies.

General discussion I

Public attitudes towards genetically modified crop plants

Elliott: Gurdev Khush, in your introductory comments you focused our attention sharply on the fact that humankind needs a dramatic enhancement of rice productivity over a rather short period, in spite of the fact that there will be an decrease in the amount of land available for cultivation coupled with a decrease in the water supply and a need to reduce the use of agrochemicals. Of course, we are dealing with a crop that has already benefited from the very best breeding technology and from huge advances in agronomy over the last several decades. It is almost an impossible target. You identified three strategies for assaulting this target, one of which was raising the yield ceiling. I wonder if you would be kind enough to say a few words about the strategies for raising the yield ceiling? In your comments you might think it appropriate to talk about gene manipulation techniques. I find it rather inspiring the way that two truly great plant breeders, yourself and Norman Borlaug, have increasingly spoken favourably about gene manipulation techniques for cereal improvement. You have both highlighted the fact that the yield ceilings of almost all of the cereals have risen very little over several decades, yet we in the West, and in Europe in particular, are confronted by what I describe as a series of 'ecoterrorist' assaults on gene manipulations. I was joking with Mike Gale yesterday about the fact that the truly great John Innes Centre is identified as the 'evil empire' by the ecoterrorists. I assume that the participants at this symposium are unified by the conviction that humankind must have the opportunity to benefit from crop gene manipulation. The success of the ecoterrorists' tactics is such that they are even having an influence on growers' strategies in the USA, something which I thought could not possibly happen. I was convinced that genetically modified (GM) products were so firmly entrenched into the cultural practices of American farmers that it was impossible to reverse their expansion. The fact that this conviction is being questioned must be a matter of concern for everyone in this room. So, what are the strategies for raising the yield ceiling, and how do we confront the ecoterrorists?

Khush: For raising the yield ceiling, we are now using different approaches, both conventional as well as those involving genetic manipulation. In the conventional approaches we are trying to modify the plant ideotypes so that we can channel more

42

energy into grain production. For doing this we have conceptualized a plant type we call 'new plant type', which some people refer to as 'super rice'. We have made considerable progress and are confident that we will have varieties with 15–20% higher yields within the next couple of years. The second approach is heterosis breeding, which is commonly applied in out-crossing species such as maize. This approach has been used successfully in China. Rice hybrids with 15–20% higher yields have been released in India, Vietnam and the Philippines. We are also trying to transfer some quantitative trait loci (QTLs) from wild species that may increase yield potential. There is evidence from Cornell that one can introgress QTLs from related species that can increase yield potential. Finally, with regard to gene manipulation we are collaborating with Tom Okita to see if we can introduce a *glgc* gene into rice which might increase the efficiency of ADP glucose pyrophosphorylase, and we have a collaborative project with your institute to introduce the isopentenyl transferase (*ipt*) gene for stay-green characteristics, which might also help us in improving the grain filling. In rice, most varieties have about 15–20% unfilled grains. If we can fill all of them, we can raise the yield potential.

With respect to the widespread opposition to GM crops, I think this is a very discouraging situation. We are all concerned about the direction the public perception of genetic engineering is going. It is disheartening to see the opposition to this elegant science that is being utilized for improving the crop yields, and particularly for helping the people in the developing countries where there is a shortage of food. Most of this opposition started from the western countries where there is no shortage of food, but they don't realize that the problems are in the developing world. They are trying to tell the developing world what to do about it, as if the people in the developing world are unable to decide what is good for themselves. Hopefully this storm will blow over, so that by the time we have the products we will be able to utilize them.

Mazur: The major agricultural companies are launching a consortium to inform the public on the benefits of the technology, the Council for Biotechnology Information. The sponsors include almost all the major companies, and was launched in April 2000. There is a web site for information (*http://www.whybiotech.com*).

Gale: I have just taken part in an inter-academy meeting, where there will be a joint statement put out by the National academies of the UK and USA, together with Brazil, Mexico, India, China, and the Third World Academy of Science. This will reiterate Gurdev Khush's last statement: that more food is needed by the developing countries and that GM is a legitimate way forward.

Dong: It seems that the agricultural companies are responding very slowly and that the damage has already been done. The *New York Times* has been carrying a full-page advertisement against GM foods every week, and there has been no response from Novartis, Monsanto and Dupont until now.

Mazur: Each of the companies has been taking a different approach. For them to form a consortium and take a common approach is not simple, but has now been accomplished.

Goff: It would probably be best if the public heard it from academic researchers. The public response stems in part from the fact that early GM products were viewed as good for farmers and not necessarily the public — the public doesn't see GM crops as offering them a big advantage. They need to realize that GM crops can be good for the environment, and that all crops consumed today have been genetically modified by standard breeding practices.

Mazur: Part of the informational programme will be to bring academic researchers into the discussion so that they may also respond to public enquiries.

Elliott: I have made a routine of accepting invitations to speak about this matter to non-specialists. I subtitle my lecture, 'A saga of misinformation, disinformation and downright poor public relations'. It seems to me extremely heartening to hear the representatives of several multinational companies agree to develop a better public relations strategy. Monsanto was the 'evil empire' even before the John Innes Centre achieved that dubious status. I note that Monsanto appreciated that what was acknowledged to be the bully-boy tactic of imposing GM soybean on people who didn't even know that they were taking it, was unfortunate. If they had actually isolated and labelled the GM material that comprised less than 3% of the first commercial harvest, I'm convinced that the ecoterrorists would have gone in a different direction and they would have been less successful. By the time Monsanto conceived what I thought was a very good advertising campaign in the UK, most people were just so convinced that Monsanto were interested only in the financial bottom line that the campaign reflected back on them. It really is an awful mess. I just hope that by the time we need the products, with luck the pendulum will have swung back.

Leach: Often when we are asked to speak on these issues, we are asked to address what research has been done on risk assessment in these areas. The problem is that we don't have access to these data for evaluation. Malcolm Elliott has said that often companies aren't trusted. If we can't have the data showing that GM crops are safe, in order to interpret them and share them with the public, it becomes a difficult issue. I understand that many data are proprietary, but you are saying trust us, but how can we?

Mazur: In order to register the products, enormous regulatory dossiers are submitted. It is all public information; it is just daunting to sort through. I believe that as part of the informational programme that I mentioned, data will be made available.

Salmeron: People are often unaware that we have active monitoring programmes for the products that are out there, which are pretty vigilant.

Leach: You are talking to the converted, but the extension people in our faculty frequently go out, and the farmers ask, 'Is this really safe? Monsanto says it is safe, but do we trust them? We want your evaluation of these data'.

Mazur: The consortium is creating a website where the relevant data will be housed.

Rice — the pivotal genome in cereal comparative genetics

Mike Gale, Graham Moore and Katrien Devos

John Innes Centre, Norwich Research Park, Colney, Norwich NR4 7UH, UK

Abstract. Over the past 15 years rice has been the focus of intense co-ordinated research activity to apply the new molecular biology to this key staple. The fact that rice has a small tractable genome and the development of genetic and genomic tools not available in any other cereal have now ensured the promotion of rice as a favoured research target. However the discovery that gene content and gene order — genome colinearity — have been maintained among all the Poaceae family for some 60 million years of evolution has elevated rice yet further to the status of a 'model' organism. Rice tools can be applied in research on the other major cereals, wheat and maize, and many aspects of rice genetics can be transferred to the many minor economic grass species that have not themselves warranted extensive research and breeding. In this paper we describe some of the applications of the discovery of extensive synteny among the grasses.

2001 Rice biotechnology: improving yield, stress tolerance and grain quality. Wiley, Chichester (Novartis Foundation Symposium 236) p 46–58

Rice genetics

Rice genetics has progressed by leaps and bounds over the past decade. The solid base laid by classical rice geneticists and breeders over the past century has been used as a launch pad for several major research initiatives. The Rockefeller Foundation's International Rice Biotechnology Program (Toenniessen 1998), started in 1986, preceded several large national programmes such as the Korean effort centred in Suwon and, most notably, the Japanese Rice Genome Program based at Tsukuba. Rice is not only an important staple food crop, it also has a much smaller genome than many other cereals including maize and wheat. As a result, rice has rapidly become the most researched cereal. Denser molecular maps and more DNA resources, including restriction fragment length polymorphism (RFLP) probes, microsatellite markers, expressed sequence tags (ESTs), bacterial artificial chromosome (BAC) and yeast artificial chromosome (YAC) libraries and genomic sequences, are available in rice than in any other cereal. The international rice genomic sequencing effort, already underway, will further consolidate the rice genome's central position.

Rice is also being used increasingly to study other larger and less tractable cereal genomes. The single, major factor that has led to the promotion of rice as a model has been the discovery of an unexpectedly high degree of similarity between the different cereal genomes in terms of gene content and gene order. It is now being recognized that conserved colinearity exists in many groups of plants, animals and microbes, but the critical observations were first made in the cereals in the late 1980s. Early reports involving rice included comparative maps of rice and maize (Ahn & Tanksley 1993) and rice and wheat (Kurata et al 1994, van Deynze et al 1995a). The genomes of these three major cereals can be shown as a single synthesis (Figure 1 and Moore et al 1995)

Other grasses of lesser global economic significance have also been shown to display similar close synteny with rice. These comparative studies include sorghum (Dufour et al 1997), sugar cane (Glaszmann et al 1997), foxtail millet (Devos et al 1998), oats (van Deynze et al 1995b), pearl millet (Devos et al 2000) and wild rice (Kennard et al 1999). Others that will follow over the next year or so include the forage grasses, fescue (O. A. Rognli, personal communication) and rye grass (J. Foster, personal communication), finger millet (M. M. Dida & K. M. Devos, personal communication) and tef (M. Sorrells, personal communication). No major surprises should be expected and it is probable that organization of all Poaceae genomes will be described by only 30 or so rice linkage blocks (Gale & Devos 1998).

Orphan crops

It is clear that the general syntenic relationships will be of immense benefit to 'orphan' crops. An example is foxtail millet, *Setaria italica*, a crop of agricultural significance only in northern China. Although foxtail millet has received very little research attention in the past, it has recently achieved a considerable genetic base by the simple strategy of having its $2n = 2\times = 18$ genome aligned with rice (Devos et al 1998). For example, QTL analyses now reveal clear alignment of major factors controlling agronomic traits with genes in wheat, maize or rice for which the physiological, biochemical and genetic control has already been well studied (Wang et al 2001).

Grass evolution

The segmental chromosome organization of the various economic grass crop species, expressed relative to the present day rice genome, can be used to track evolutionary relationships. One early and clear result is that maize is exposed as

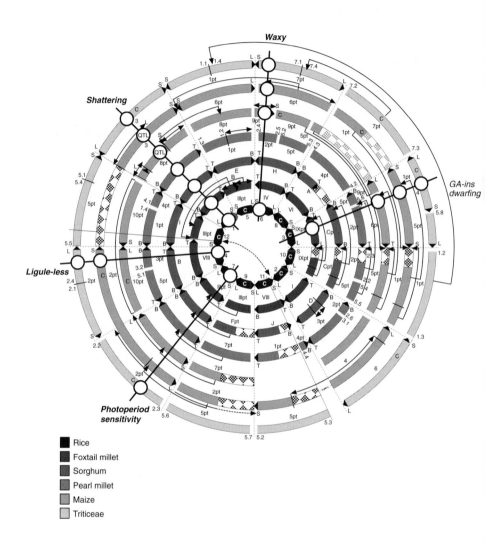

FIG. 1. Comparative maps of several of the world's cereal crops species genomes. The 12 chromosomes of rice, ten chromosomes of maize and the basic seven chromosomes of the Triticeae, together with the genomes of pearl millet, sorghum and foxtail millet, are aligned so that homoeologous (orthologous) loci lie on the same radius. The arrangement is simply the most parsimonious, requiring the fewest number of adjustments from the present day chromosome linear gene orders, and is not intended to represent any genome as the most primitive. Evolutionary translocations and inversions, relative to rice that have to be invoked are shown by the arrows. Chromosomal regions where the syntenic relationships are still not clear are left blank. For the additional detail see text.

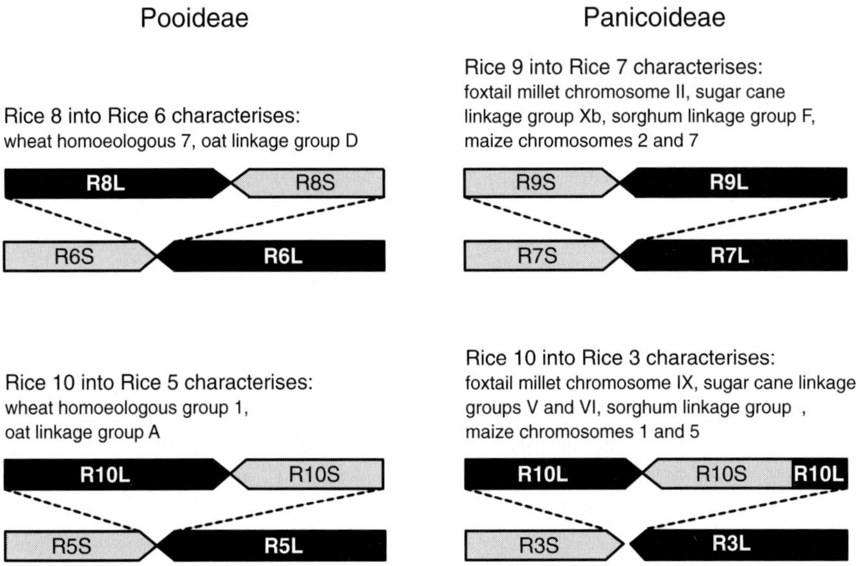

Pooideae

Panicoideae

Rice 9 into Rice 7 characterises:
foxtail millet chromosome II, sugar cane
linkage group Xb, sorghum linkage group F,
maize chromosomes 2 and 7

Rice 8 into Rice 6 characterises:
wheat homoeologous 7, oat linkage group D

Rice 10 into Rice 5 characterises:
wheat homoeologous group 1,
oat linkage group A

Rice 10 into Rice 3 characterises:
foxtail millet chromosome IX, sugar cane linkage
groups V and VI, sorghum linkage group ,
maize chromosomes 1 and 5

FIG. 2.   Evolutionary translocations in the grasses.

an almost complete tetraploid (Fig. 1). Other major rearrangements, dating back 60 million years, can be identified that define the Pooideae, as exemplified by oats and wheat, and the Panicoideae, as exemplified by pearl and foxtail millet, sorghum, maize and sugar cane. These can be seen in Fig. 1. For example, in the Pooideae, insertion of present day rice chromosome 10 (R10) into R5 represents the structural organisation of wheat chromosome 1 and oats chromosome A. Interestingly, in these latter two species, the region corresponding to rice chromosome 10, which displays normal recombination with a map length of 150 cM in rice, is highly compressed to just a few map units in the low recombinogenic centromeric regions of the cereals with larger genomes. This and other key changes that define the major grass groups are shown in Fig. 2.

Questions that cereal taxonomists will probably debate for many years to come are, 'what was the primeval grass genome?', and 'how many chromosomes did it have?' One small step towards resolving the structure of the ancestral genome may be provided by the comparative organisation of the Pooideae and Panicoideae chromosomes. The involvement of R10 in two different insertion events would suggest that an independent R10, i.e. as in rice itself, is the more primitive configuration.

It has been suggested that the extent of evolutionary chromosomal rearrangement is a reflection of evolutionary time (Paterson et al 1996). However, care must be taken with this line of thinking because these rearrangements can clearly become fixed at varying rates in different species. For example, the rearrangements between relatively closely related Triticeae genomes, for example rye (Devos et al 1993) or *A egilops umbellulata* (Zhang et al 1998) relative to wheat, are almost as extensive as those between wheat and rice.

The degree to which segmental chromosome duplication that can be expected in rice, and in the other cereals, is also an important question. Extensive duplication will be a hindrance to applications such as cross genome map-based cloning where 'walks' could get deflected by inadvertent jumping from one segment to its duplicate. In addition, duplicated genes may be more prone to divergence and/or silencing leading to deletion which will result in apparent disruption of colinearity. Information from other species is beginning to indicate that duplication is far more common than previously thought. For example, *Arabidopsis*, the 'simple' model diploid, turns out to be more than 60% duplicated (Mayer et al 1999). Rice is almost certainly no exception although there is not enough genomic sequence data yet available to derive an equivalent statistic. However information concerning ancient large duplications in the grasses is beginning to emerge. The R11/R12 duplication, which involves substantial portions of the short arms of both chromosomes, was originally thought to be specific to rice (Nagamura et al 1995). However the same duplication has now been shown to exist in pearl and foxtail millet (Fig. 3). Thus this duplication clearly predates the divergence of the Bambusoideae and Panicoideae subfamilies.

Undoubtedly more duplications will be identified which, while being potentially useful predictors for the taxonomists, will confound the work of those wanting to exploit the conservation of gene order between genomes.

Gene prediction

At the gross map level, and considering genes of major effect, the predictive power of the alignments holds up well. Figure 1 shows the comparative locations of a few genes. These include major structural loci, such as the waxy genes, and classical morphological mutants, such as genes for liguleless and dwarfism. Often, major genes in one species are aligned with QTLs for the same trait in another, as is the case with the domestication related shattering genes in wheat and rice which are aligned with QTLs for the same trait in both maize genomes.

Cross genome gene isolation

Cereal genome alignments can be used not only to predict the presence of genes but also as a tool to isolate genes. Until recently wheat genes were well beyond the reach of conventional map-based cloning technology because the genomic tools were not available and because of the large genome size. The homoeologues of the

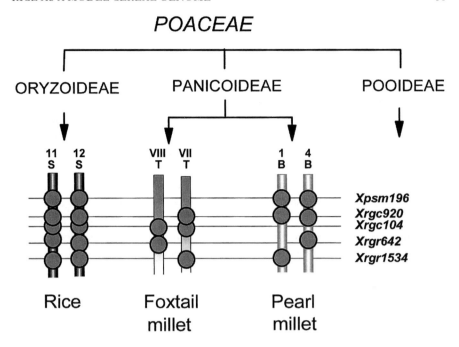

FIG. 3.   The R11/R12 duplication in rice, foxtail and pearl millet. The figure shows that RFLP markers that define the rice duplication are similarly duplicated and map in the same linear orders in the other two species.

same genes in rice, however, are far more tractable to isolation. The tools — very dense genetic maps, large EST collections, large-insert BAC, PAC or YAC libraries — are available and the total genome size, at 430 Mb, is far less daunting. This approach is being used to isolate the wheat *Ph* gene, which controls chromosome pairing by limiting it to homologues, rather than homoeologues (Riley & Chapman 1964). Detailed maps of the critical regions on wheat chromosome 5B and rice chromosome 9 show a very high level of conserved colinearity (Foote et al 1997). Because allelic variation is not known at the locus in wheat, let alone rice, a new approach has been developed to narrow down the region that contains *Ph* (Roberts et al 1999) in which a set of overlapping deletions, induced by fast neutron mutagenesis, were used rather than recombination. The 300 kb region of the rice YAC contig corresponding to the section of wheat 5B containing *Ph* has been sequenced and the painstaking functional genomics exercise to identify the roles of the 20 or so open reading frames in the region has begun (G. Moore, personal communication).

This strategy should be applicable to any gene in any crop. Allelic variation is not a prerequisite and large recombinant populations are not necessary. As the rice genomic sequence in its entirety becomes available this may become a more popular approach.

Monocot–eudicot synteny?

The close alignment of genes between genomes separated by 60 million years of evolution begs the question as to the extent of synteny remaining over the 140–200 million year monocot–eudicot divide. Significant synteny would mean that many cereal genomics approaches could be serviced by the *Arabidopsis* DNA sequence that has just become available.

This question has been addressed by several studies (Paterson et al 1996, Devos et al 1999, van Dodeweerd et al 1999) and the consensus is that, although there is some evidence of residual synteny in some regions of the genome, gene sequences and orders have diverged so much that there is very little predictive power of comparisons. The recent revelations that extensive chromosome segment duplication are present in *Arabidopsis*, and the suggestion that it might even have an allo-polyploidy event in its past, would seem to definitely preclude the use of arabidopsis gene orders to approach cereal genetics.

This is indeed rice's time.

References

Ahn SN, Tanksley SD 1993 Comparative linkage maps of the rice and maize genomes. Proc Nat Acad Sci USA 90:7980–7984

Devos KM, Atkinson MD, Chinoy CN et al 1993 Chromosomal rearrangements in the rye genome relative to that of wheat. Theor Appl Genet 85:673–680

Devos KM, Wang ZM, Beales J, Sasaki T, Gale MD 1998 Comparative genetic maps of foxtail millet (*Setaria italica*) and rice (*Oryza sativa*). Theor Appl Genet 96:63–68

Devos KM, Beales J, Nagamura Y, Sasaki T 1999 Arabidopsis-rice: will colinearity allow gene prediction across the eudicot–monocot divide? Genome Res 9:825–829

Devos KM, Pittaway TS, Reynolds A, Gale MD 2000 Comparative mapping reveals a complex relationship between the pearl millet genome and those of foxtail millet and rice. Theor Appl Genet 100:190–198

Dufour M, Deu M, Grivet L et al 1997 Construction of a composite sorghum genome map and comparison with sugarcane, a related complex polyploid. Theor Appl Genet 94:409–418

Foote T, Roberts M, Kurata N, Sasaki T, Moore G 1997 Detailed comparative mapping of cereal chromosome regions corresponding to the Ph1 locus in wheat. Genetics 147:801–807

Gale MD, Devos KM 1998 Plant comparative genetics after 10 years. Science 282:656–659

Glaszmann JC, Dufour P, Grivet L et al 1997 Comparative genome analysis between several tropical grasses. Euphytica 96:13–21

Kennard W, Phillips R, Porter R, Grombacher A, Phillips RL 1999 A comparative map of wild rice (*Zizania palustris* L. 2n = 2x = 30). Theor Appl Genet 99:793–799

Kurata N, Moore G, Nagamura Y et al 1994 Conservation of genome structure between rice and wheat. Bio-Technology (NY) 12:276–278

Mayer K, Schüller C, Wambutt R et al 1999 Sequence and analysis of chromosome 4 of the plant *Arabidopsis thaliana*. Nature 402:769–777

Moore G, Devos KM, Wang ZM, Gale MD 1995 Cereal genome evolution. Grasses, line up and
    form a circle. Current Biol 5:737–739
Nagamura Y, Inoue T, Antonio BA et al 1995 Conservation of duplicated segments between rice
    chromosomes 11 and 12. Breed Sci 45:373–376
Paterson AH, Lan TH, Reischmann KP et al 1996 Toward a unified a genetic map of higher
    plants transcending the monocot–dicot divergence. Nat Genet 14:380–382 (erratum: 1997
    Nat Genet 15:322)
Riley R, Chapman V 1964 Cytological determination of the homoeology of chromosomes of
    Triticum aestivum. Nature 203:156–158
Roberts MA, Reader SM, Dalgliesh C et al 1999 Induction and characterization of Ph1 wheat
    mutants. Genetics 153:1909–1918
Toenniessen GH 1998 Rice biotechnology capacity building in Asia. In: Ives CL, Bedford BM
    (eds) Agricultural Biotechnology in International Development. CAB International,
    Wallingford, p 201–212
Van Deynze AE, Nelson JC, Yglesias ES et al 1995a Comparative mapping in grasses. Wheat
    relationships. Mol Gen Genet 248:744–754
Van Deynze AE, Nelson JC, O'Donoughue LS et al 1995b Comparative mapping in grasses. Oat
    relationships. Mol Gen Genet 249:349–356
van Dodeweerd AM, Hall AM, Bent EG, Johnson SJ, Bevan MW, Bancroft I 1999
    Identification and analysis of homoeologous segments of the genomes of rice and
    Arabidopsis thaliana. Genome 42:887–892
Wang ZM, Thierry A, Panaud O, Gale MD, Sarr A, Devos KM 2001 Trait mapping in foxtail
    millet. Theor Appl Genet, in press
Zhang H, Jia J, Gale MD, Devos KM 1998 Relationships between the chromosomes of *Aegilops
    umbellulata* and wheat. Theor Appl Genet 96:69–75

DISCUSSION

Mazur: You were very emphatic in your statements, but I know that you are
aware of Renato Tarchini's work (Tarchini et al 2000) showing that there can be
interruptions at the microsynteny level. When you sequence maize and rice you
find interruptions of the synteny, yet you stated that every gene will be syntenous.

Gale: The sequence that I was talking about on rice chromosome 9 does look to
be particularly good. There are complications there; there are duplications within
that as well. Jeff Bennetzen's group have published some work on a fully sequenced
sorghum BAC and a rice BAC. In this case, every gene was there. It is true that at
Dupont you looked at a sequence region of rice and then looked at the relationship
with corn by hybridization, because you didn't actually have the corn sequence. It
looked as though some genes were missing. This is very interesting. Could it be
that if they are missing, this is because maize is an ancient tetraploid, and therefore it
has been able to lose one or other of the two copies of these genes?

Mazur: There were some that were missing, but there was one that was on a
different chromosome all together.

Dong: I don't remember the phylogenetic tree, but do you think it is better to
sequence the most primitive grass in evolutionary terms rather than the later ones?

Gale: That is a nice thought, but which is the most primitive grass species? I guess we don't know. Although rice is one of the smaller grass genomes, it is by no means the smallest. In fact, foxtail millet, *Setaria italica*, has a genome the same size as rice.

Presting: If I understood you correctly, then wheat chromosome 1 is composed of rice chromosome 10 flanked by rice 5 at both ends. Foxtail millet has a rice chromosome in which rice chromosome 10 is flanked by rice 3. What does this say about the evolution of these chromosomes?

Gale: You could argue that because rice 10 appears in two different chromosomal environments in wheat and foxtail millet, the rice configuration, with 10 as an entity on its own, is the most primitive. There is a lot more work to be done in this area.

Mazur: I was surprised by the absence of polymorphism in the wheat chromosomes. Is this due to the lack of transposons in wheat? Why is it so different from what I would expect in corn?

Gale: If John Doebley (White & Doebley 1998) is right, the corn genomes have been out there, collecting mutations, for 21 million years. The wheat chromosomes have been out there for only 10 000 years. I believe that wheat is of monophyletic origin. That is, bread wheat is derived from a single cross between the tetraploid and *Triticum tauschii*. If this did happen just once, then 10 000 years ago wheat was effectively a single doubled haploid. So all the polymorphism that we have in wheat has to have arisen over just the last 10 000 years, relative to maize's 21 million. It is not surprising that wheat is relatively unpolymorphic. The polymorphism in *T. urartu* and *T. tauschii* today is vastly greater than in the A and D genomes of wheat. Remember the genetics: there is one pollen grain, which just contains one copy of each of the chromosomes, fusing with one egg. No matter how polymorphic the individual species was, you would only get one chromosome haplotype coming across on the day bread wheat was created.

Mazur: You are looking at A, B and D and not seeing any polymorphism between the three of them.

Gale: That is not quite true. We have done some sequence work on this (Bryan et al 1999), and our story is that within a single gene sequence, we expect one base pair in a thousand, on average, to distinguish varieties — and these can be very different varieties indeed. Between the same locus over the three genomes, it is one base pair in a hundred. This is an order of magnitude difference, which may be just a million years' worth of evolution.

Khush: Now we know that maize is also an ancient tetraploid. Is there any similar mechanism like the *Ph* gene in wheat, to suppress the homeologous pairing?

Gale: That is a key question. We only know about *Ph* because Ralph Riley (1959) and Okamoto (1962) were able to observe its effect by removing the chromosome, 5B, which carries the gene. Effectively, we had a knockout genotype in which to

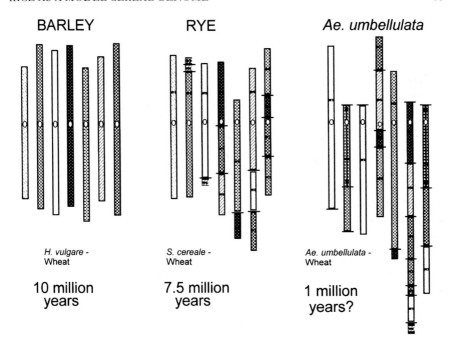

FIG. 1.   *(Gale)* The figure shows the rearrangements in barley, rye and *Ae. umbellulata*, relative to wheat. The barley genome is identical with hexaploid bread wheat (*Triticum aestivum*).

look at the effect. As far as I am aware no one has yet observed a similar phenotype in corn.

G.-L. Wang: Since rice and other cereal crops have been separated by 60 million years, there must be some evolutionary force acting to maintain the colinearity among the cereal genomes. What kind of mechanism is responsible for this?

Gale: That is a good question, and one that we don't have the answer to yet. In fact the situation is quite complicated. If you look at Fig. 1 [*Gale*] you will see three *Triticeae* genomes, each described in terms of the present day wheat genome.

The barley genome has been separated from wheat by ∼ 10 million years, yet the genome is almost indistinguishable from that of wheat. Rye, at ∼ 7.5 million years, is younger than barley but the chromosomes are considerably rearranged relative to wheat. Within the separate chromosome segments colinearity is near perfect. At the time we said that this difference was probably because barley was an inbreeder and rye was an outbreeder (Devos et al 1993). However, we have recently mapped *Ae. umbellulata* relative to wheat (Zhang et al 1998). This close relative of wheat is considerably younger than rye and it is an inbreeder, however it is more rearranged relative to wheat than even rye. It is clear that the rate at which evolutionary chromosome translocations are accumulated is not a function simply of elapsed evolutionary time. Something else is driving these changes and, at the moment, it is not clear what this force is.

Khush: When you are talking about rye, presumably you are talking about *Secale cereale* or cultivated rye. Cultivated rye differs from wild species in two to three translocations, and this could probably account at least partly for these rearrangements.

Gale: It certainly describes the rearrangements, but I'm not sure it accounts for them. You are quite right. In order to derive the present day *S. cereale* genome, starting with present day wheat, you need to invoke a minimum of seven translocation events. The first four of these would bring you to the *Secale montanum* genome (Devos et al 1993). This would suggest that *S. cereale* is more distant from wheat. These are diploids, so each time they make a translocation they will have to go through one generation where they are semi-steriles. Yet we have managed to fix all of those translocations.

Leung: In the case of disease resistance genes, is it true that they don't follow the colinearity as well as other structural genes? For example, in rice, it was impossible to get the homologue for the barley rust resistance gene.

Gale: This is a strange story. This is a piece of work being done in Andy Kleinhoff's laboratory in Pullman, where they were looking for a rust resistance gene, *Rgp1*, which matches with the top of rice chromosome 1. They have really gone to town on this piece of chromosome and have pulled out some 60 genes, almost all of which have counterparts in rye and in barley, except *Rgp1*. The last time I talked to Kleinhoff, he said they had one candidate left to try, but it didn't look like it was a resistance gene (Han et al 2000). There probably is good reason to suspect that disease resistance genes might evolve faster than other genes and may therefore lose their integrity in this sort of analysis.

Pryor: In fact the maize *Rp1-D* gene maps close to the barley *Rpg1* (Ayliffe et al 2000). There is something of a dispute at the moment as to whether it actually cosegregates with it. There is the problem of identifying specificity among a small gene family. The tip of chromosome 10, from your original maps, was an anomaly, where you weren't able to put it into your maps with rice and wheat. This was the region for the *Rp1-D* locus. This may support Hei Leung's idea: the regions of chromosomes where some of these resistance loci are found appear to be outside some of the generalizations.

Mazur: The genes that didn't hybridize to maize were disease resistance genes in that maize/rice colinearity study.

Goff: Genes under strong selective pressure might vary from this synteny model, not just disease resistance genes.

Gale: To be sure, when you look at straightforward single copy genes, such as the isozyme loci and major morphological mutants, the model works fine. But to take it to the next step we need more examples like Jeff Bennetzen's work (Chen et al 1997) where there are complete sequences on both sides.

Leung: What is the reason for the lack of colinearity with microsatellite markers?

Gale: Now that people are pulling microsatellites out of ESTs you may see a different story, but the original cereal genome microsatellites were produced from enriched genomic libraries, so we are almost always looking outside of genes. Colinearity refers only to the genes themselves. Clearly there are many differences in DNA sequence between genomes in the inter-genic regions, and this means that PCR markers using primer sequences from within these regions will tend to be genome specific.

Bennett: Can you comment on the expansion of the size of the cereal genomes? What mechanisms are operating to ensure expansion is uniform along the chromosome that the synteny doesn't jig, thus allowing you to draw nice straight lines between the genomes?

Gale: Another piece of work from Jeff Bennetzen's group (SanMiguel et al 1996) has revealed maize transposons that transpose only into themselves. These sorts of structures were simply never observed while people were looking only at gene sequences. It was only when they started to sequence through an entire YAC clone that they began to see them. An active transposon that inserts only into itself will allow rapid build up of genomic sequence that will, effectively, push the genes apart, but will not disrupt gene order. Moreover, if you take the point of view that recombination takes place mainly in genes then these rafts of transposons are not going to be reflected in the genetic map. Genes will look as if they are next door to one another but, in reality, will be separated by long lengths of repeats. As yet this mechanism has been observed only in maize, and similar situations are not yet, to my knowledge, becoming apparent from the genomic sequence available in rice or other species.

Mazur: An important part of this is that when the elements transpose into themselves they are not interrupting the genes.

Goff: A recent publication comparing genome sizes in *Drosophila* species reported that certain *Drosophila* species with small genomes had the ability to eliminate repetitive DNA more efficiently. Is there any evidence that rice is better at eliminating or preventing amplification of repetitive DNA versus species with larger genomes such as maize?

Gale: Not that I am aware of. However, it is interesting: if rice has the same number of genes in 400 Mb that wheat has in 16 billion base pairs, how can wheat afford the energy cost of replicating all that DNA with each cell division? One would think that small is beautiful in terms of efficiency of replication. We always think in terms of genomes getting bigger with evolution, but there is no reason why they shouldn't also get smaller.

Salmeron: If you look at rearrangements across dicot species of similar times of divergence, how does it compare to the monocot cereals?

Gale: In general, the relationships across the grasses appear among the most conserved. However, studies in the eu-dicots are revealing similar situations. The 10 million years of crucifer evolution that have separated the *Brassica* crop species from *Arabidopsis* have left very strong colinearity. Legume genomes appear to be particularly rearranged, however colinearity within lineage blocks can be observed. Similar good relationships are emerging from studies of forest trees. I would say that synteny will prove to be the rule rather than the exception.

References

Ayliffe MA, Collins NC, Ellis JG, Pryor A 2000 The maize *rp1* rust resistance gene identifies homologues in barley that have been subjected to diversifying selection. Theor Appl Genet 100:1144–1154

Bryan GJ, Stephenson P, Collins AJ, Smith JB, Gale MD 1999 Low levels of DNA sequence variation among adapted genotypes of hexaploid wheat. Theor Appl Genet 99:192–198

Chen M, SanMiguel P, Bennetzen JL et al 1997 Microcolinearity in *sh2*-homologous regions of the maize, rice and sorghum genomes. Proc Natl Acad Sci USA 94:3431–3435

Devos KM, Atkinson MD, Chinoy CN et al 1993 Chromosomal rearrangements in the rye genome relative to that of wheat. Theor Appl Genet 85:673–680

Han F, Chen JP, Kudrna D et al 2000 Sequence analysis of a rice BAC covering the syntenous barley *Rgp1* region. Genome 42:1071–1076

Okamoto M 1962 Identification of the chromosomes of common wheat belonging to the A and B genomes. Can J Genet Cytol 4:31–37

Riley R, Chapman V, Kimber G 1959 Genetic control of chromosome pairing in intergeneric hybrids in wheat. Nature 183:1244–1246

SanMiguel P, Tikhonov A, Bennetzen JL et al 1996 Nested retrotransposons in the intergenic regions of the maize genome. Science 274:765–768

Tarchini R, Biddle P, Wineland R, Tingey S, Rafalski A 2000 The complete sequence of 340 kb of DNA around the rice *Adh1-Adh2* region reveals interrupted colinearity with maize chromosome 4. Plant Cell 12:381–391

White S, Doebley J 1998 Of genes and genomes and the origin of maize. Trends Genet 14:327–332

Zhang H, Jia J, Gale MD, Devos KM 1998 Relationship between the genomes of *Aegilops umbellulata* and wheat. Theor Appl Genet 96:69–75

Bioinformatics for rice resources

Bruno W. S. Sobral, Harry Mangalam*, Adam Siepel*, Pedro Mendes, Rob Pecherer* and Graham McLaren†[1]

*Virginia Bioinformatics Institute, Virginia Tech (0477), 1750 Kraft Drive, Suite 1400, Blacksburg VA 24061, USA, *National Center for Genome Resources, 2935 Rodeo Park Drive East, Santa Fe, NM 87505, USA, and †International Rice Research Institute, DAPO Box 7777, Metro Manila, Philippines*

Abstract. The distinguishing feature of the 'new biology' is that it is information intensive. Not only does it demand access to and assimilation of vast data sets accumulated by engineered laboratory processes, but it also demands a previously unimaginable level of data integration across data types and sources. There are various information resources available for rice. In addition, there are various information resources that are not focused on rice but that contain rice data. The challenge for rice researchers and breeders is to access this wealth of data meaningfully. This challenge will grow significantly as international efforts aimed at sequencing the entire rice genome come into full swing. Only through concerted efforts in bioinformatics will the power of these public data be brought to bear on the needs of rice researchers and breeders worldwide. These efforts will need to focus on two large but distinct areas: (1) development of an effective bioinformatics infrastructure (hardware systems, software systems, and software engineers and support staff) and (2) computational biology research in visualization and analysis of very large, complex data sets, such as those that will be developed using high-throughput expression technologies, large-scale insertional mutagenesis, and biochemical profiling of various types. In the midst of the large flow of high-throughput data that the international rice genome sequencing efforts will produce, it is also imperative that integration of those data with unique germplasm data held in trust by the CGIAR be a part of the informatics infrastructure. This paper will focus on the state of rice information resources, the needs of the rice community, and some proposed bioinformatics activities to support these needs.

2001 Rice biotechnology: improving yield, stress tolerance and grain quality. Wiley, Chichester (Novartis Foundation Symposium 236) p 59–84

The term bioinformatics may have originated about 100 years ago in Germany. At that time, the context was the pursuit of a mathematical description of life. The term now has broader and more varied meaning and context in the modern age of genomics. It can be said that modern bioinformatics is concerned with all

---

[1]This chapter was presented at the symposium by Graham McLaren, to whom correspondence should be addressed.

aspects of acquisition, processing, storage, distribution, analysis and interpretation of biological information. It combines the tools and techniques of mathematics, statistics, computer science, communications and biology with the aim of distilling biological knowledge from a variety of data sources.

According to this broad definition, bioinformatics has been around for a long time. The acquisition, utilization and publication of traditional data types such as phenotypic evaluation, adaptation, classical genetics and socioeconomics has been an integral part of agricultural research for the past century. However, the new aspect of bioinformatics, and the one that most people take subconsciously as a definition, concerns the integration of modern information technology and the 'new' (largely reductionist) biology — molecular data, genome sequence data and proteomics.

One powerful key to unlocking the secrets of the 'new' biology, however, is the functional integration of traditional and new data sources, as these provide the necessary context for utilizing molecular information. The opportunity is that developments in information technology now make this integration possible and bioinformatics is bringing it about.

Biology has been enriched many times through the migration of scientists from other disciplines. Through physicists, biology took on a reductionist approach that has resulted in great progress in the new domain of molecular biology. Engineers have recently entered biological laboratories and helped to convert them into high-throughput data factories or 'data farms'. Computer scientists and software engineers have contributed through the creation of biological databases. As a result of these various changes, we now live in a world where a single information resource can serve thousands of geographically distributed users, thus justifying the expense in creation and maintenance of such systems.

Because of changes in biological research, the information overload of our times has definitively reached biological researchers worldwide. For example, a search on the term '*Oryza sativa*' gave more than 4000 results in December 1999 using Google as a search engine. Over 9000 'hits' were returned using the term 'rice genome' and over 3000 using 'rice germplasm'. The problem does not stop with rice, of course, since rice researchers need to access information from other model species as well as from multi-species repositories, such as DNA sequence databases (for example, the Genome Sequence DataBase, GSDB, *http://www.ncgr.org/research/sequence/*), protein databases (for example, SwissProt *http://www.expasy.ch/sprot/*), gene expression resources (for example, GeneX *http://www.ncgr.org/research/genex/*), and biochemical pathway resources (such as KEGG *http://www.genome.ad.jp/kegg/kegg2.html* and PathDB *http://www.ncgr.org/research/pathdb/*). Clearly, navigation of this information has become a major chore for rice and other researchers.

Not only are the types of information resources changing from mostly libraries to electronic repositories, but it is also clear that the very nature of scientific publishing in biology is undergoing a change. For example, when the genome of *Arabidopsis thaliana* is completed in 2000, there will be publications describing the results. However, it is very likely that these papers will contain little of the raw data from the ∼ 150 Mb genome of *A. thaliana* and focus on highlights that the authors consider of relevance. Thus, the only repository that will contain almost all of the relevant public data and thus be the most useful to a broad range of scientists interested in *A. thaliana* or other organisms will be 'The *Arabidopsis thaliana* Information Resource' (TAIR, *http://www.arabidopsis.org*). In addition, TAIR is experimenting with novel methods for sharing of private data sets with public researchers, such as the recent example with the Cereon polymorphism data (*http://www.arabidopsis.org/cereon/index.html*). In the future, scientific publications may become more electronic; an experiment currently addressing changes in publication of biological papers is the *Journal of Agricultural Genomics* (JAG, *http://www.ncgr.org/research/jag/*), where authors have an opportunity to link their research publications to the integrated data they have analysed and interpreted, thus allowing other investigators to access and interpret the same data sets with other assumptions in mind.

Despite wonderfully rich information resources for rice (described below), supported by the communities they serve, there is a need for integration of these resources at a higher level. Integration would to allow any investigator with Internet access to have a complete view of the model system without needing to navigate all these sites and integrate the information manually. Such has been the experience of the human genome, yeast genome (*http://genome-www.stanford.edu/saccharomyces/*) and *Arabidopsis* genome (*http://www.arabidopsis.org*) communities. It makes sense for the rice-science communities worldwide to learn from the lessons of the communities that have undergone the 'genomic shift'.

Integration of molecular information (such as molecular genetic linkage and physical maps, biochemical pathways, DNA sequences, protein information, and gene expression data) is important, but it is even more important (and challenging) to integrate these molecular data sets with organismic data, typically provided through germplasm repositories and mutant stock centres. This need is clear because organisms are more than the sum of their parts and because it is the organism-level data that provide context to the molecular data. Unfortunately, there has been less development of germplasm information management systems (GIMS) than of laboratory information management systems (LIMS); thus basic informatics infrastructure is missing. However, an interesting prototype for a GIMS has been studied and implemented by members of the Consultative Group on International Agricultural Research (CGIAR) — it is known as the International Crop Information System (ICIS, Fox & Skovmand 1996, *http://www.cgiar.org/icis/homepagetext.htm*). Preliminary implementations have occurred for rice, wheat, and maize at the International

Rice Research Institute (IRRI) and the Centro Internacional de Mejoramiento de Maiz y Trigo (CIMMYT).

Once integration of molecular information with phenotypic information is achieved in a robust manner, it is very likely that the resulting system will modify the way we think about breeding. Typically, plant breeding is done by crossing and selecting desirable individuals from the progeny. With the opportunity to make predictions concerning the outcomes and to model *in silico* the desired genotype environment combinations to optimize selection for specific traits, breeding shifts in its character. One of the changes is that breeders start to become model testers themselves and model systems (or other information-rich systems) become useful tools for model building. Once models of phenotype×genotype× environment are verified through explicit breeding experiments, the task becomes one of moving the models themselves around through breeding in different organisms. One very interesting effort that is pursuing this type of paradigm shift, is the Quantitative Genetics (QU-Gene, *http://pig.ag.uq.edu.au/ qu-gene*) simulation system.

There is an overarching need also to think about the processes that all this information needs to support, from the perspective of making genomic information useful to those needing to use rice as a crop species. Paramount among these users are breeders in developing and developed nations (*http:// www.ncgr.org/bioinformatics/paper/*, Sobral et al 2001). Those users are not going to benefit from bioinformatics developed to support drug discovery in pharmaceutical companies — specific effort and resources are needed to make this happen.

Finally, once integration from molecules to phenotypes has occurred, the next frontier becomes integration of ecosystems information (environment and interactions among organisms). There are various efforts attempting to achieve this integration in specific cases. One such effort is represented by FloraMap (*http://www.floramap-ciat.org*), a computer tool for predicting the distribution of plants and other organisms in the wild.

Current state of rice information resources

Organismal information

RiceWeb (*http://www.riceweb.org*) is a portal to some rice information resources and has been developed jointly by three centres from the CGIAR, namely IRRI, the Centro Internacional de Agricultura Tropical (CIAT) and the West African Rice Development Association (WARDA). A list of rice databases maintained by international rice research centres is given at *http://www.riceweb.org/database.htm* but many of these are not currently available online. These information resources typically contain organismal or environmental data, although some isozyme data are included as well.

Germplasm information resources for rice include the International Rice Genebank Collection Information System (IRGCIS). Developed to manage the genebank collection at IRRI, it manages information on all aspects of germplasm management and use—passport data, characterization and evaluation data, and germplasm inventory. It is also used to manage operations for the collection such as generating field books, choosing germplasm for rejuvenation and tracking where germplasm has been sent under obligations to FAO. Currently IRGCIS is accessible only through IRRI's LAN.

The System-wide Information Network for Genetic Resources (SINGER, *http://www.cgiar.org/singer*) is curated by the International Plant Genetic Resources Institute (IPGRI) in Rome and attempts to link information on all genetic resources managed by the CGIAR. Rice data from IRGCIS are replicated to SINGER periodically. You can find information on passport and some characterization data. SINGER Phase II has been initiated, with the goal of developing a more user-friendly interface, and permitting access to more data. In fact, through SINGER, in 2000 it will be possible to provide web access to IRGCIS. Thus the database will be searchable interactively and germplasm can even be requested electronically.

IRRI has also developed a number of small genebank systems for national programs in Laos, Myanmar, Bangladesh, India (for rice) and Malaysia, based on IRGCIS in terms of structure and use, but for use on a PC with Microsoft Access.

The International Network for Genetic Evaluation of Rice (INGER) is a network for the exchange and multi-location testing of elite/improved germplasm. INGER was founded in 1975 as the International Rice Testing Program, and it permits breeders to determine the value of their lines for release as varieties and to select lines for use in hybridization. IRRI has more than 20 years' data from these trials. IRRI has developed INGERIS in a RDBMS. The International Rice Information System (IRIS) is the rice implementation of ICIS, and information from IRGCIS, INGERIS and all plant breeding and evaluation projects of IRRI are being integrated through it. IRIS is also being extended to IRRI's partners in rice research so that new data are immediately integrated with historical data. This bi-directional access is available by CD-ROM, and read only access is being implemented through the Web.

Rice as a model for genomics of cereals

In the new era of plant genomics, rice has sprung up as a major model for the Poaceae (grasses or cereals), and thus an international effort has been assembled to sequence the entire genome of rice. A newsletter for the international consortium for the sequencing of the rice genome can be found at *http://www.staff.or.jp/rgp/News/Newsletter.html*. Some highlights of information resources for genomic activities in rice follow. The description is not intended to be exhaustive; rather it is illustrative of the diversity of groups and approaches that are being taken.

The Korean Rice Genome program presents its genomic data at *http://202.30.98.73/ricemac.html*. This information resource shows the current status of the international effort to sequence the genome of rice (*http://202.30.98.73/ricegenome/status.html*). For graphical views of sequencing results, this resource links back to GenBank (for example, for chromosome 1, we have *http://www.ncbi.nlm.nih.gov/cgi-bin/Entrez/framik?gi =5922603&db=Nucleotide*). Graphical chromosome views of rice at the National Center for Biotechnology Information (NCBI) can be seen at *http://www.ncbi.nlm.nih.gov/PMGifs/Genomes/new_euk_g.html*. FASTA format for DNA sequences is also available through the Korean resource. Rice linkage map information is displayed at *http://202.30.98.88/krgrp/Map.qry?function=form*, from where it is possible to access any of the following maps: Academia Sinica Map; Cornell Map; Japan NIAR-STAFF Tsukuba Map; Korea Map — NIAST, RDA; and Morphological Map, as searchable entities. A graphical display of the Korean linkage map is shown at *http://202.30.98.73/transgenic/krgrpmap.gif*. Expressed sequence tags (ESTs) can be found at *http://202.30.98.88/krgrp/RiceEST.qry?function=form*, where they can be searched. Transgenic Rice in Progress at Myongji BioScience is also shown on the site.

For bacterial artificial chromosome (BAC) end-sequencing of rice, there is a link to the Clemson University project at *http://www.genome.clemson.edu/projects/rice.html*. At the Clemson BAC centre's site, BAC end sequences can be downloaded in bulk or searched as a database at *http://www.genome.clemson.edu/projects/rice_bac_end.html*. BLAST searches can be performed on the searchable database at *http://www.genome.clemson.edu/bacdb/htdocs/index.html*.

In Japan, the Rice Genome Project (RGP) is a joint project of the National Institute of Agrobiological Resources (NIAR), the Institute of the Society for Techno-innovation of Agriculture, Forestry and Fisheries (STAFF) and a part of the Japanese Ministry of Agriculture, Forestry and Fisheries (MAFF) Genome Research Program. The RGP site is at *http://www.staff.or.jp/*. Within the RGP site, INE (pronounced i-ne) is the Japanese word for rice plant. It is used here as an abbreviation for INtegrated rice genome Explorer — a database integrating the genetic map, physical map and sequencing information of rice genome. This database is written in Java. Access to genomic sequencing data is at *http://www.staff.or.jp/GenomeSeq.html*. A description of the international rice genome sequencing collaboration and roles is at *http://www.staff.or.jp/Seqcollab.html*.

The USDA–ARS-funded RiceGenes resource is located at *http://genome.cornell.edu/rice/*. There has been a new release reflecting a large increase in the amount of sequence data, which will be updated on a monthly basis, with the new data available near the 1st of each month. In December 1999, the database contained the following: 9 genetic maps of rice; 3 comparative maps of rice (with maize, oat and the Tritiaceae); 19 840 probes; 86 176 sequences; 26 264 *Oryza* sequences; 59 912 BLAST hits on the *Oryza* sequences; approximately 457 quantitative trait loci (QTLs); 342 rice QTLs; 115 maize QTLs; approximately 1500 references; and 100 rice variety releases. From here it is possible to access the Cornell and Japanese versions of rice linkage maps or to download the current RiceGenes database. Microsatelitte data are at *http://genome.cornell.edu/rice/microsats.html*.

Cold Spring Harbor's sequencing effort is shown at *http://nucleus.cshl.org/riceweb/*, where BLAST searches can be conducted and preliminary annotation is shown (for an example see *http://nucleus.cshl.org/riceweb/10P20-titlepage.html*). A local version of RiceGenes is also presented at *http://stein.cshl.org/perl/ace/search/ricegenes*.

The Institute for Genomic Research's (TIGR) rice gene index (an EST alter ego) is at *http://www.tigr.org/tdb/ogi/*. Rutgers' Plant Genome Initiative site shows their progress on rice genomic sequencing at *http://pgir.rutgers.edu/Riceprogress/Island1progress.html*.

The National Center for Gene Research of the Chinese Academy of sciences has its progress on rice genomics at *http://www.ncgr.ac.cn/index.html*. Following the completion in 1996 of a BAC contig map of the rice Indica genome, the Center is now focusing on the large-scale DNA sequencing of chromosomes with the intention to improve rice breeding.

The Taiwanese rice genome efforts are shown at *http://biometrics.sinica.edu.tw/genome/index_e.htm*. The Taiwanese effort is interested in collaborating closely with laboratories involved in the International Rice Genome Sequencing Project, where Taiwan proposes to sequence chromosome 5. One central laboratory in the Institute of Botany, Academia Sinica is conducting a pilot sequencing study. Graphical representation of the progress made on chromosome 5 is presented at *http://biometrics.sinica.edu.tw/genome/PAC.htm*.

Genoscope's participation in rice genome sequencing is shown at *http://www.genoscope.cns.fr/externe/English/Projets/Projet_CC/CC.html*. As part of a collaborative project together with the CNRS and University of Perpignan , the CIRAD and the IRD (ex ORSTOM), Genoscope is sequencing a 1 Mb region localized on chromosome 12 of *Oryza sativa* ssp. japonica cv. Nippobare/GA3. This region is close to a major locus conferring resistance to the rice yellow mottle virus.

Information resource needs and technologies

Bioinformatics involves continued research in the areas of information theory, data visualization and algorithm development. This research component of bioinformatics is well suited to an academic or research-driven funding and deployment model. However, there is a second major area of bioinformatics that provides methods to acquire, store, analyse and distribute biological data. That component can be seen as infrastructural and is typically the domain of engineers. It has been recently recognized by the NIH in the Biomedical Information Science and Technology Initiative (BISTI), a US interagency working group (*http://www.nih.gov/welcome/director/060399.htm*), that 21st century biologists will need:

- Programs of excellence in specific topics (the initiative suggests the creation of 20 bioinformatics centres in the US non-profit and academic institutions).
- A program for information storage, curation, analysis and retrieval (ISCAR).
- Adequate resources for computational support at the single principal investigator grant level.
- A national computer infrastructure that can be scaled with the growth of data.

It is clear in reading the document that there is an understanding of the need for bioinformatics infrastructure as well as the research components of bioinformatics. Additionally, NSF and NIH have been exploring a partnership in computational biology, focused around the area of supercomputing.

Biological databases, the Internet and genomics

Two major developments have contributed to the plethora of biological databases that now exist. The first was the appearance of the biologically oriented database in the late 1970s (Dayhoff et al 1980). There are now more than 4 billion nucleic acid and protein residues in over 64 databases undergoing a continuing exponential growth in size (Benson et al 1999). The second computer science advance was the advent of packet-switched computer networks resulting in the Internet. The Internet provides for expansion of biological database availability to a broader community worldwide. Now a single resource could service thousands of researchers thereby justifying investing considerable effort into making common information resources. The Internet also allowed widespread collaboration on projects that did not require physical proximity so that projects that had no hope of succeeding with one developer could garner enough distributed support to succeed, such as was the case with open source (OSrc) projects such as Linux (*http://www.linux.org/*) and ACeDB (*http://www.sanger.ac.uk/Software/Acedb/*), many genomic sequencing projects, and the world wide web itself.

There are currently many rice information resources, reviewed briefly already. Diverse rice information resources provide for varied needs of different communities. However, this diversity and lack of coordination of many critical resources can severely limit, if not impede, the effective utilization of the results of public investments in rice research worldwide. Additionally, there is a deluge of molecular data coming from plant genomics that needs to be dealt with quickly. Thus, we need to preserve the diversity of approaches and information resources for rice; and at the same time provide the rice community with informatics infrastructure and methods for integration of rice data.

In developing information resources for rice, our target users are primarily researchers from academia, non-profit organizations and government. Large agrochemical companies can pay large sums to implement custom solutions, but most of our target users do not have the funding to acquire a commercial package or a custom local solution. This blocks access to the very people who might be able to use the data most creatively!

'Factory' biological data, produced in engineered laboratories, whether composed of DNA sequences, mRNA expression profiles or metabolite profiling results, has some basic common requirements from the perspective of informatics. It is always necessary to provide methods, hardware and software infrastructure for acquisition, storage, querying, analysis and visualization of the massive volumes of data. Typically, these data are referred to as 'high-throughput' (HT) data and they result from our increasing capacity to measure and characterize cellular components in living cells.

It is possible to think about HT data as being of various 'types'. These types reflect the research world in terms of specialties and funding. For example, DNA sequences (or protein sequences) have a well-defined format based on the genetic code. It is easy to see that the properties of other types of data, such as patterns of mRNA expression over time in specific tissues, or metabolic pathways, are not of the same nature and thus have distinct requirements. However, it is necessary to be able to move easily from DNA sequences to biochemical pathways to study the behaviour of genomes with respect to changing environmental stimuli. This necessity seems obvious when we consider that in living organisms the HT data types represent integrated components.

Integrating and distributing key rice data types

Considering the importance of rice as a staple crop worldwide and its emerging role as a model system for the understanding of Poaceae in general (the major economically relevant plant group), it is necessary to provide integrated sites for the following types of rice data:

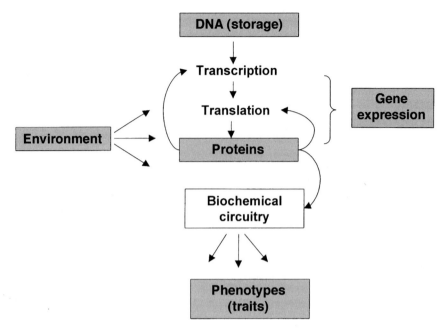

FIG. 1.   Reverse engineering of complex adaptive systems, from molecules to phenotypes.

- Structural genomics: DNA sequences (to complete genomes) and maps
- Gene expression: mRNA profiling, northerns
- Biochemistry: pathways (metabolic and signalling), metabolites, proteomics
- Germplasm (phenotypic data): natural variants and induced mutant collections with their respective phenotypic characterization — closely tied to environmental and geographical data as appropriate.

Structural genomics.   The data types above progress from the purely molecular at the top toward the expression of phenotype at the bottom (as shown in Fig. 1). For each of these types of data it will be necessary to provide the means for acquisition, storage, querying, analysis and visualization. This is best handled through generalized 'components'. Because of the unity of life and life processes it is easier to solve the informatics need for each data type and then use the general solution across multiple organisms to enable the power of comparisons. This is most famously illustrated through the use of similarity searches of DNA sequence databases using tools such as BLAST (*http://seqsim.ncgr.org/newBlast.html*). Sequence databases such as GSDB or EMBL can be seen as warehouses that contain all public DNA sequences from all organisms. These can be thought of as being one 'component' or subsystem within a whole that integrates the different components through some architecture. That a warehouse may be updated from various distributed, specialized databases is particularly encouraging, especially if communication standards pave the way.

Following this notion of a system with specialized components, the bulleted points above correspond to subsystems that can be built from generalized components and populated with rice (or any other) data. Data population and curation should be as painless as information technology and evolutionary (iterative) software development can make it. Processes should be set up and maintained by rice researchers. Ongoing data curation and editing must be seriously considered and resources allocated at the outset. For example, with DNA sequences it would be optimal to provide methods for semi-automated data acquisition and preliminary analysis. One example of such a system is the *Phytophthora* Genome Initiative (*http://www.ncgr.org/research/pgi/*), where simple tasks are performed automatically for geographically distributed researchers collaborating on the study of a biological process.

Additionally, gene families can be of particular interest to biologists and yield important basic and practical information through their study. While large warehouses provide a place for all public data to reside, it is important that highly curated data sets be developed through analysis by experts in the field, and the results provided freely to the public research community. An example of such a dataset is the Plant Disease Resistance Genes Database (*http://www.ncgr.org/research/rgenes/*). Similar data sets could be produced for key rice gene families, or plant gene families as compared to say rice, *Medicago truncatula* and *A. thaliana*, thus providing insight into the three main groups of flowering plants, from a social and economic perspective.

As a given genome is studied with HT methods, maps become useful viewing paradigms for genome-scale information, especially for genetic applications. An important goal of genome mapping efforts is to contribute to the efficiency and effectiveness of using similarities and differences between and among genomes and genome maps as a path to new knowledge. This is especially critical for crops that are not the object of industrial-scale investment by life sciences companies. Because of the distributed nature of public plant genome efforts, data import from heterogeneous sources is not a one-shot problem, yet it is required for meaningful integration and maximum return on investments in plant biology. All active genome projects are continually updating their databases. Individual communities operate under independent update cycles. Maintaining up-to-date copies of genome maps from external sources is another challenge, especially since the same maps are often available from several sources. Which graphic objects appear in a map visualization depend on the user's needs. A phylogenetic approach might focus on entire regions of synteny, while the search for genes would be enhanced by sequence data and gene prediction software. Whatever the application, the paradigm of a genome map appears to be one effective means for visualizing and navigating the relevant data. All these map comparisons have one thing in common, namely the detection and display of relationships between map objects based on the map structure. However, the ability to 'connect' mapped objects is not limited to structural relationships based on common probes. It is also possible to detect relationships *in silico* based on non-map structural similarities such as sequence similarity, or functional 'relatedness' as correlation in gene expression experiments. Of course, all *in silico* discovery is hypothetical so results will inevitably require confirmation in the laboratory!

Gene expression.   Genomics has allowed the sequencing of complete genomes. These data have permitted researchers to analyse what all the genes in a particular organism are doing at one time (Phimister 1999). Methods for studying large-scale gene expression, while unparalleled in power, also promise to increase data flow by orders of magnitude, unleashing a tsunami of gene expression data.

The increase in production of gene expression data has not been matched by corresponding development in software tools and databases to analyse and exploit the results of such experiments. Generating large-scale gene expression data is currently so expensive that many investigators are not budgeting adequately for controls to validate their data. Also, techniques for examining overall gene expression are so new that even the technology vendors do not yet understand all the complexities involved, so many of the datasets that have been generated are not being analysed correctly or completely.

Gene expression data sets tend to be larger than sequence entries and only a fraction of data produced is of direct interest to the group generating the data, thus the extraction of information tends to be sparse. Other groups should be able to glean this under-utilized information by querying public gene expression systems.

There is an inclination to treat gene expression data in terms of biological sequence: as a much larger set of data, but approachable in the same way with the same basic set of tools. This is incorrect. While some of the tools (BLAST, Altschul et al 1990), Hidden Markov Models (Krogh et al 1994), and regular expressions (Friedl 1997) have applications to gene expression, the data are not only different in kind, but much more complex. In a sequence database, the core data are strings of nucleic (or amino) acid residues. There may be variation of a given sequence across a population, but within an individual, the sequence is invariant. In contrast, gene expression is highly dynamic and it is this variation in the expression patterns which researchers will use to attribute relationships between genes. Another difference between gene expression data and other biological data is that it contains little inherent information; the value or meaning of gene expression data comes from the context of the experiment. What was the taxonomy, sex, and developmental stage of the organism? What were its growth conditions? From what organ and tissue was the sample extracted? What protocols were used in sample preparation? To be able to answer these questions, given the data complexity, it is critical to start as soon as possible to design interfaces and systems that can be used effectively and efficiently by biologists to gain better understanding of the multiple interrelationships among genes in these data. It will take several iterations of design, implementation, feedback, and redesign to provide an intuitive tool useful to a biologist. Meanwhile, there is a concurrent explosive growth in data volume compounded by rapidly developing laboratory technology.

To enable efficient searching across the wide range of experimental conditions in expression databases, it will be important to develop restricted vocabularies to replace the use of free text. However, the problems associated with establishing satisfactory nomenclatures, especially across species, are substantial (Gelbart 1998). Even in taxonomy, where there has been an established formalism for centuries, there is disagreement on vocabulary. The situation is much less settled for such recent and dynamic domains as developmental biology, plant and animal pathologies, gene and protein naming conventions, metabolic relationships, or laboratory protocols, all of which will play a major role in determining the utility of gene expression systems. There is a need for communities to collaborate and share restricted vocabularies.

To handle mRNA profiling experiments, the National Center for Genome Resources (NCGR) is collaboratively developing an integrated toolset that will allow researchers to upload, browse, query, analyse and visualize results of mRNA profiling experiments. This subsytem will support most microarray technologies as well as others for which there are substantial data streams, such as Amplified Fragment Length Polymorphisms (AFLP, Vos et al 1995) and Serial Analysis of Gene Expression (SAGE, Velculescu et al 1995). GeneX (*http://www.ncgr.org/research/genex/whitepaper.html#design*) includes an RDBMS, an extensible schema that encodes the necessary relationships among the data and meta-data, the scripts and applications that handle the data input/output, the interface code through which the user interacts with the RDBMS, and the analytical applications. It also includes the Java code for a client-side Curation Tool that is used to annotate and validate the data before it is uploaded to a database. Initially this tool will be used to communicate to distributed GeneX databases, but when a common Gene Expression Markup Language (GEML) format has been defined, it can be used to communicate to any database that can parse this format. NCGR is using the OSrc software development model, so that we can incorporate other such tools and to ensure that the complete system can be distributed, encouraging others to develop their own analytical and visualization routines. The basic design will be scalable to handle the range of needs from lab data production environments to large public data repositories. The advantages of such a software resource include:

- A freely available, extensible gene expression database amenable to lab use, encouraging data producers to use both the Curation Tool and the database to store and analyse their own data (this also encourages distributed data curation, relieving a serious bottleneck of providing basic description of the data in a manner compatible with a relational database).
- Easier, more robust analysis of primary data, leading to faster publication and more efficient contribution of the data to a public repository.
- Schema coherence with higher level GeneX databases.
- Compatibility with any GEML-compatible database.

The advantages of such a public repository of gene expression data include:

- Widespread availability of data and therefore improved, distributed analysis of gene expression data including better error correction, statistical approaches, data mining and visualization techniques, especially for researchers either unable to generate their own gene expression data or more interested in theoretical problems concerning the data.
- Comparison of data from different technologies, allowing the results from one technology to be more accurately compared to that of another. Currently, this is discouraged by many of the technology vendors.
- Collections of gene expression profiles under base conditions in different organisms to provide better statistical bases for measuring other conditions.
- Accessibility of gene expression data from refereed journals and their public examination and validation.

The GeneX system will be a stand-alone relational database system coupled to tools that allow filtering and clustering using various parameters and algorithms. To this end a few original tools (tacg, *http://24.1.175.29/tacg*, overrep, Cyber-T) have been developed and other freely available ones have been incorporated.

Biochemistry.   The availability of complete genomes, starting with prokaryotes, has also allowed metabolic reconstruction based on identified genes, enzymes, and known pathways. This is currently achieved through use of annotation (meta-data) of DNA sequence data from a given genome. With the nearing completion of the *A. thaliana* genome, scheduled for release in mid-2000, metabolic reconstruction of *A. thaliana* based on this approach is now possible (Waugh et al

2000). While the reconstruction of biochemistry in any organism is possible through meta-data associated with DNA sequences, the future holds even more promise. Rather than electronic repositories of data that can be found in books, exciting initiatives are focused at metabolic reconstruction, simulation and understanding, based on data acquired from experiments measuring metabolites. A quantitative focus is emerging in biochemical bioinformatics. One notable effort is PathDB (*http://www.ncgr.org/software/pathdb/*), where the main data types represented are compounds, reactions, enzymes and other metabolic proteins and pathways. PathDB attempts to include very rich descriptions of the kinetic, thermodynamic and physicochemical properties of pathway components. Currently, PathDB is focused on acquiring and curating the last 70 years of biochemical knowledge from the literature, in an organism-prioritized manner. However, unlike other existing metabolic databases, PathDB is not restricted to the pathways that the curators have defined (the ones commonly found in textbooks). PathDB allows for any set of connected reactions to be considered a pathway. This allows users to pose questions such as: 'What are all the pathways that connect compound A with compound B?', essentially pathway discovery. Pathway discovery starts to allow metabolic engineers wishing to change the biochemical characteristics of an organism to do so explicitly, using bioinformatics to make predictions about performance. How the organism is engineered (traditional breeding vs. genetic engineering) is a separate factor as biochemical engineering can be achieved by either approach.

If PathDB is an example of a pathway information resource that is aimed at enabling quantitative analysis and discovery of pathways, then the main associated tool that can use the data to support metabolic engineering approaches and 'what if' analyses on pathways is the General Pathways Simulator (GEPASI, *http://www.gepasi.org*). GEPASI is a freely available program for the simulation and optimization of the kinetics of systems of chemical and biochemical reactions (steady states and time courses). Currently, GEPASI runs in Microsoft Windows 95 and NT 3.51 and above, both on Intel and DEC Alpha platforms. Models in GEPASI v. 3 can be composed of many compartments with different volumes. The number of reactions and metabolites in each model is only limited by available memory. Simulations can be followed interactively (including adding perturbations to a time course). GEPASI characterizes steady states using metabolic control analysis and linear kinetic stability analysis. It provides a means for advanced exploration of a model's behaviour in multi-dimensional parameter space. It is capable of finding maxima or minima of any model variables with any number of adjustable model parameters, and of data fitting (parameter estimation) with experimental data. GEPASI's data output is flexible and can be inputted into most post processing programs (spreadsheets, graphics programs, artificial neural networks, etc.). The results of simulations can be plotted in 2D and 3D directly

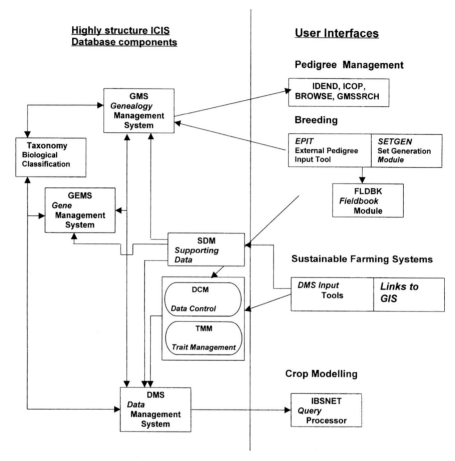

FIG. 2. The International Crop Information System. Integrated data components and specialized user interfaces.

from the program (GEPASI uses gnuplot, also an open source package — a major advantage of the OSrc model is being able to incorporate other OSrc software).

Phenotypic data. As previously mentioned, The International Rice Information System is one implementation of ICIS, a database system for the management and integration of global information on genetic resources and crop improvement for any crop (Fig. 2). It is being developed by genetic resources specialists, crop scientists and information technicians in several of the centres of the CGIAR and in the national systems (NARS). ICIS provides a comprehensive crop information system implemented separately for each crop and is based on unique identification of germplasm, management of nomenclature, and retention of all pedigree information. Collaborative software development is possible through adoption of an open programming environment, allowing access to different database systems through the Open Database Connectivity protocol (ODBC), and facilitating multi-language programming through Dynamic Link Libraries (DLLs). Internationalization of information systems is facilitated through wide area networking, or on CD-ROM. At the same time, it allows capture of new pedigrees or information on mutant line development, into a private local database, which can be exported periodically to the central database at the user's discretion. Developments are already underway to achieve access through the web based on protocols developed through SINGER.

ICIS (IRIS) permits scattered information generated on germplasm to be integrated, linked to sources of seed, and put to work for plant improvement, or other research applications in genomics and molecular biology, for instance. Its distributed design allows all partners to have access to the latest information. It has two main components. The first is the Genealogy Management System (GMS) to:

- Assign and maintain unique germplasm identification.
- Retain and manage information on genealogy.
- Manage nomenclature and chronology of germplasm development.

The second main component is the Data Management System (DMS) to:

- Store and manage documented and structured data from genetic resources, variety evaluation, and crop improvement studies.
- Link data to specialized data sources such as GMS, location, and climate databases.
- Facilitate queries, searches, and data extraction across studies according to structured criteria for data selection.

Previous attempts at integrated data management systems for agricultural research have foundered due to confusion between data storage and visualization and analysis. Scientists from different disciplines want to see and manage data in different ways, and this has always been allowed to dictate the data model for storage. ICIS overcomes this constraint by using a single, general data model but allowing the development of specialized user interfaces giving distinct views of the integrated data (Fig. 1). Interfaces currently being developed are for plant breeding, crop modelling and farming systems. Others are planned for genetic resources management, molecular biology and modelling of genetic systems.

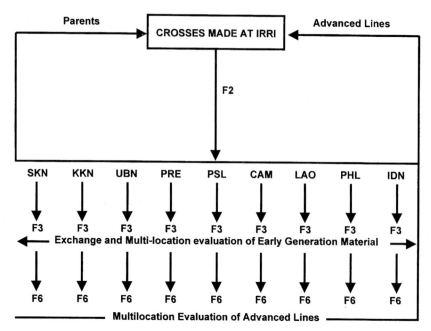

FIG. 3.   Rainfed lowland shuttle breeding network for South East Asia. CAM, Cambodia; IDN, Indonesia; KKN, Khonkaen, Thailand; LAO, Laos; PHL, Philippines; PRE, Phrae, Thailand; PSL, Phitsanulok, Thailand; SKN, Sakhon Nakhon, Thailand; UBN, Ubon Ratchatani, Thailand.

One of the first disciplines which needs to tap into an integrated data management system is crop improvement, and the ICIS breeders' interface allows dispersed breeding programs such as IRRI's SouthEast Asian Shuttle Breeding Network (Sarkarung et al 1995) to exchange information on pedigree evaluations at the same time as exchanging segregating material (Fig. 3). This allows an understanding of plant adaptation and genotype by environment interaction can be brought to bear on each crop improvement cycle.

Systems integration and rapid evolutionary software development

Systems integration

Phenotypic performance is the consequence of the integration of processes at all levels of the biological system — molecular, biochemical and ecological. In order to understand or model phenotypes it is essential that we integrate data across the compatible subsystems. Various approaches to deliver this requirement to scientists can be taken and some have already yielded interesting results. Some of

the most prominent efforts have come in the form of systems to enable querying across heterogeneous databases, and in the promotion of component-based software development. In addition, some groups have built integrated systems, according to different models.

Most efforts to enable cross-database querying address the problem of data retrieval independently from and to the exclusion of analysis and visualization. Some systems use analytical search tools such as BLAST in a way that is largely analogous to querying a database, and some allow for generic visualization (i.e. visualization that treats different types of database entities as fundamentally the same). Although they differ in significant ways, Kleisli (Chung & Wong 1999, Davidson et al 1999, *http://www.cbil.upenn.edu:8089/K2/k2web?page=home*), TAMBIS (Baker et al 1998, 1999) and OPM (Markowitz et al 1997, 1999, *http:// gizmo.lbl.gov/opm.html*) are representative of this general approach. Kleisli is notable for its powerful functional query language, CPL, and its sophisticated optimization strategy. TAMBIS ambitiously attempts to define a global 'ontology' (also called a 'knowledge base' — Karp 1996) for biological concepts, and for its knowledge-driven graphical query interface that enables users a high degree of insulation from source databases. OPM provides a global object-oriented data model, and well-developed tools for editing, querying, and browsing database contents. All of these efforts take unmediated multi-database or federated-database strategies (Karp 1996), choosing to query source databases 'on the fly', rather than to replicate their data in a data warehouse.

Improved software interoperability through component-based design is largely in its infancy in bioinformatics but there have been a few notable efforts. Advocates of component-based design argue for a strategy in which various groups create software in the form of independently deployable building-blocks, or components, which adhere to standards sufficient to ensure that they can be assembled variously into larger systems. This approach reduces duplication of effort, enables greater participation by small informatics groups, facilitates rapid prototyping and software maintenance, insulates groups from the complexities of one another's software, and enables 'visual programming' using Integrated Development Environments (Goodman et al 1996, Boyle 1998, *http:// www.cbil.upenn.edu/bioWidgets/*).

Some groups (Boyle 1998, *http://www.cbil.upenn.edu/bioWidgets/*) have focused their development on visualization components, without much attention to the questions of where the data or analysis services to supply these data are to come from, and how they are to be acquired and integrated. In this way, their work falls to the opposite extreme of those focusing on cross-database querying. Other groups, most notably the Object management Group's (OMG) (*http:// www.omg.org*) Life Sciences Research Task Force (*http://www.omg.org/about/ old%20report.htm#lifesciences*), advocate a component-based approach to data

retrieval, in which autonomous databases present standard interfaces on the Internet that any other software component can use to retrieve data adhering to a standard object-oriented data model. Essentially, the case for components is compelling and generally unobjectionable; the question is how to make it happen. The OMG's LSR Task Force is making the largest-scale and steadiest progress towards component-based software in bioinformatics, but its consensus-building standardization process, while probably a sensible approach, is slow and laborious, and it will take considerable time before the necessary standards are not only established but widely enough in use for programmers to have substantial options for 'plug-and-play' software development.

One common and simple approach to data integration, to hypertext-link networks of HTML pages (which are often dynamically generated by CGI scripts or servlets) using the World Wide Web, represents a simplified kind of component-based design. One web page points to another by way of a URL, which is a component interface in that it represents a contract to show some information or provide some service, without any promise of how such responsibilities are to be met. Like other component-based approaches, systems of HTML pages can be built from the bottom-up, and are forgiving of heterogeneity. The 'interfaces' between HTML pages, however, are limited in the richness of their semantics, and after one 'component' (i.e. web page) invokes another, HTML and browser technology, at least in its current state, does not really allow ongoing interactions between them. Furthermore, browser-based user interfaces are noticeably discontinuous, requiring an expensive server call to process each set of user inputs, and consequently can be slow to respond. Though HTML browser-based approaches can still be cost effective — when one considers the increased development time and installation complications of the alternatives — they are limited, and in the long run may be destined for displacement by systems that take better advantages of the powerful computers on our desktops.

There are a few other bioinformatics systems that bear mentioning here. The ACeDB system (*http://www.sanger.ac.uk/Software/Acedb/*) represents an early, ambitious attempt to provide biologically meaningful visualization, along with browsing and querying capabilities, for data of a variety of different types and sometimes spanning multiple species. It is widely used, especially in the agricultural community, and as an open-source project, has an enthusiastic and public-spirited development community.

Ritter (1994) talked about an 'Integrated Genomic Database (IGD)', which may be the earliest reference to an attempt to develop an integrated system for genomics that meets our notions of integration. IGD approached both the problems of integrating data of different types that comes from multiple, heterogeneous repositories, and of providing an intuitive, graphical interface to allow browsing, querying, and visualization of those data. It used a data warehousing approach and

employed ACeDB tools for visualization and browsing. IGD included map, sequence, phenotype, and population data, and focused on the human genome. Although it was implemented successfully as a prototype, development seems to have been discontinued.

Finally, several companies have built 'next-generation' integrated systems for their own commercial use or for sale to others, using promising new technologies such as Java and CORBA to achieve platform-neutral, distributed deployments, and using component-based architectures for maximum flexibility, extensibility, and re-usability. These systems, while powerful, appear to be prohibitively expensive, and inaccessible to the majority of not-for-profit researchers.

NCGR is taking a different approach to the problem of integration: a generalized platform for the integration of heterogeneous bioinformatics resources is under construction (*http://www.ncgr.org/research/isys/*). Using the platform, separately developed components behave as an Integrated System (ISYS). The platform employs an open, extensible, component-based architecture, enables event-based communication between components, and allows the system to take on different global properties based on the particular set of components present in any given configuration.

Rapid evolutionary software development

Many of the problems in bioinformatics point toward the practices of rapid evolutionary development (RED) and component-based design, which are gaining currency in software development circles. Rapid evolutionary development advocates the refinement of rapidly-developed prototype software into systems for deployment in short, iterative cycles, and is particularly well-suited for projects in volatile fields. The essential idea of RED is that most software systems are too complex, and their requirements too nebulous, to 'get right the first time', as engineers have been trained to do for mechanical and physical systems. According to this philosophy, better metaphors for software development can be found in biological systems and living organisms than in construction or manufacturing (Arthur 1992). This is particularly compelling for the difficult sociology of getting professional software engineers and biologists to communicate about their needs and systems.

Component-based design has many important benefits, both in general and in bioinformatics in particular, and combines well with RED. In particular, RED can be used simultaneously at the level of individual components and at the level of systems of multiple components. The existence of components speeds prototyping of new systems because often they can be built wholly or partly from existing components. Components are similar to objects in that they

interact according to well-defined interfaces, which behave as contracts for functionality, but they differ from objects in that they can be deployed independently, and are more easily subject to composition by third parties (Szyperski 1998). As we have noted, widespread adoption of component-based development in bioinformatics, among other things, could encourage software reuse, improve maintainability, and insulate developers from one another's complexities; these things imply decreased development time and higher quality software.

Acknowledgements

Special thanks to Michael Jackson (IRRI) for personal communications concerning rice germplasm information resources.

References

Altschul SF, Gish W, Miller W, Myers EW, Lipman DJ 1990 Basic local alignment search tool. J Mol Biol 215:403–410

Arthur LJ 1992 Rapid evolutionary development: requirements, prototyping and software creation. John Wiley & Sons, New York

Baker PG, Brass A, Bechhofer S, Goble C, Paton N, Stevens R 1998 TAMBIS: Transparent Access to Multiple Bioinformatics Information Sources. In 6th International Conference on Intelligent Systems for Molecular Biology, Montreal, Canada. AAAI Press, Menlo Park

Baker PG, Goble C, Bechhofer S, Patton NW, Stevens R, Brass A 1999 An ontology for bioinformatics applications. Bioinformatics 15:510–520

Benson DA, Boguski MS, Lipman DJ et al 1999 Genbank. Nucleic Acids Res 27:12–17

Boyle J 1998 Building component software for the biological sciences. CCP11 Newsletter, 4:22, 14. *http://www.hgmp.mrc.ac.uk/CCP11/newsletter/vol2.2/john/index.html*

Chung SY, Wong L 1999 Kleisli: a new tool for data integration in biology. Trends Biotechnol 17:351–355

Davidson SB, Buneman OP, Crabtree J, Tannen V, Overton GC, Wong L 1999 BioKleisli: integrating biomedical data and analysis packages. In: Letovsky S (ed) Bioinformatics: databases and systems. Kluwer Academic Publishers, Norwell, MA, p 201–211

Dayhoff MO, Schwartz RM, Chen HR, Hunt LT, Barker WC, Orcutt BC 1980 Nucleic acid sequence bank. Science 209:1182

Fox PN, Skovmand B 1996 The International Crop Information System (ICIS): connects Genebank to breeder to farmers' field. In: Cooper M, Hammer GL (eds) Plant adaptation and crop improvement. IRRI/CABI, Wallington, p 317–326

Friedl JEF 1997 Mastering regular expressions: powerful techniques for Perl and other tools. O'Reilly, Cambridge

Goodman N, Rozen S, Stein L 1995 The Importance of standards and componentry in meeting the genome informatics challenges of the next five years. In: Second Meeting on the Interconnection of Molecular Biology Databases. *http://www.ai.sri.com/~pkarp/mimbd/95/abstracts.html*

Karp P 1996 A strategy for database interoperation. J Comp Biol 2:573–586

Krogh A, Brown M, Mian IS, Sjolander K, Haussler D 1994 Hidden Markov models in computational biology. Applications to protein modeling. J Mol Biol 235:1501–1531

Markowitz VM, Chen IMA, Kosky A 1997 Exploring heterogeneous molecular biology databases in the context of the object-protocol model. In: Suhai S (ed) Theoretical and computational methods in genome research. Plenum Press, New York, p 161–176

Markowitz VM, Chen IMA, Kosky A, Szeto E 1999 OPM: object–protocol model data management tools '97. In: Letovsky S (ed) Bioinformatics: databases and systems. Kluwer Academic Publishers, Norwell, MA, p 187–199

Phimister B 1999 Going global. Nat Genet (suppl) 21:1

Ritter O 1994 The integrated genomic database. In: Suhai S (ed) Computational methods in genome research. Plenum Press, New York, p 57–73

Sarkarung S, Singh ON, Roy JK, Vanavichit A, Bhekasut P 1995 Breeding strategies for rainfed lowland ecosystem. In: Fragile lines in fragile ecosystems. Proceedings of the International Rice Research Conference. IRRI, Los Baños, Philippines, p 709–720

Sobral BWS, Waugh M, Beavis W 2001 Information systems approaches to support discovery in agricultural genomics. In: Phillips RL, Vasil IK (eds) Advances in cellular and molecular biology of plants, vol I: DNA-based markers in plants. In press

Szyperski C 1998 Component Software: beyond object-oriented programming. Addison Wesley Longman Limited, Essex

Velculescu VE, Zhang L, Vogelstein B, Kinzler KW 1995 Serial analysis of gene expression. Science 270:484–487

Vos P, Hogers R, Bleeker M et al 1995 AFLP: a new technique for DNA fingerprinting. Nucleic Acids Res 23:4407–4414

Waugh M, Hraber PT, Weller J et al 2000 The Phytophthora genome initiative database: informatics and analysis for distributed pathogenomic research. Nucleic Acids Res 28:87–90

DISCUSSION

Goff: The sequence information in the public databases is growing fast. Genbank has grown in an exponential fashion. There are now almost 2 million EST sequences from humans. Several of the more important crop plants have jumped from several thousand to 50 000 ESTs. Synteny information, mapping information and allelic diversity are also rapidly growing. Expression profiling and proteomics data, metabolite profiling and phenotypic profiling are additional growing data sets. Genomics techniques generate large data sets, and it is important to have bioinformatics to simplify data analysis. Biologists have to become better computer scientists. Physicists and chemists have already become better computer scientists, and this is what biology is now faced with. We have to deal with complex data sets in a way that we didn't have to five or 10 years ago.

Leach: Earlier we heard that Japan has a database on rice sequencing, the companies all have databases and there is a database starting at IRRI in collaboration with several other organizations. How close are we to having a nice database where everyone has access to each piece of information?

Leung: Our effort at IRRI is not to create something independently, but whatever platform we make here is intended to be exportable or directly integrated into the international databases.

Leach: My question is, are all of these groups talking to one another, so that the data is transferable between one database and another? This is not a trivial problem.

McLaren: This is an objective of NCGR, who are an interesting group in the sense that they don't have any data of their own. They are built up of a team of biologists and computer scientists working together — if they don't have access to other people's data they don't have any. Their objective is this integration of different data types. They have designed an interesting concept, which they call their software bus. This links applications from different organizations, or their own applications, into this bus, which is an 'intelligent' linkage system. One application will ask the bus if it has any information on maps, for example. Another will send this information down. This is the first attempt at linking diverse data types. Our initial project will be to try to link our ICIS database into this bus as one of the applications that will talk to it. If this works well, almost any other analysis application or database can be linked in and will provide the integration you are asking for without requiring everyone to get together and agree to develop together.

Gale: Will it actually work in practice? It seems to be axiomatic that bioinformaticists are most effective when they are located where the data is being produced. Certainly this is the case in the plant community in the UK. In the US, USDA completely wound up the Washington operation recently and shipped it to Cornell, which is where the data were being produced. Is an operation like the one in New Mexico going to work? Will people willingly supply them with the data that they need? Moreover, will they understand the data that has been produced by others? I recently saw a list that had eight wheat ESTs on it. It may well be that there are just eight wheat ESTs, but the people who actually work with wheat know that there are hundreds of sequenced cDNAs available. It is just that we may not have called them ESTs, and therefore the people who make EST databases haven't picked them up.

McLaren: I think it can work. Your wheat cDNA/EST dichotomy is an example where somebody pulling all these things into a data warehouse would be able to find both types of data and put them in the same place. Whether the politics of data ownership will allow it to happen is a different matter.

Goff: From a company perspective, I would like to see certain bioinformatics tools viewed as precompetitive and developed in the public domain. It is clear that we have underestimated the demand for computer analysis in genomics, and bioinformatics is often rate-limiting when this many data are available. There is also very little academic bioinformatics training at present. Many of the academic bioinformatics experts have been recruited into industry, and there is currently a shortage of training programs.

Mazur: The amount of data that are being generated, particularly as people go to chip-based technologies and arrays, is overwhelming. The bioinformatic needs are enormous. You can't retrospectively create new databases because it is too expensive to move existing data. You have to be able to link the data somehow to make it more useful.

Leach: Are the databases that are being designed such that I can compare data from database A with database B, to find out where overlap exists?

Mazur: That is the trend of the future, but currently I wouldn't say this exists. The other problem is that people are very set in their ways: if they have been trained to use one database, they don't want to be trained on another user interface.

Goff: It is pretty hard to design a perfect database in a situation where everything is changing so fast. By the time you come up with what you think is a perfect database for all the data and technologies today, in a year's time everything will be different and your database will be obsolete. It is necessary to design with change in mind.

Gale: Is the presence of the companies inhibitory to all of this? We would like to produce these nice databases, but is there a thought in the back of peoples' minds that the multinationals are simply waiting for the public sector to do this and will then just take it on board? Is there something that the companies can do to make it a more level playing field? It is admirable that the Japanese at the Rice Genome Project in Tsukuba are putting the rice sequence in the public domain, but why should they? So that US multinationals can simply suck it up and do what they want with it?

Goff: One point made this morning is that Novartis did fund significant research in Rod Wing's lab at Clemson that is now in the public domain. This is the rice BAC end sequencing and fingerprinting. When Novartis has a goal that is compatible with public release, this is a favourable way to go. Companies can support public projects in a proactive fashion.

Mazur: That is a good point. It is one thing to prospectively do that; it is another thing to retrospectively put something in the public domain after you have spent a lot of money to gain competitive advantage that way. Many of the companies would have preferred that the public sector had taken it on initially and not have this public/private wall. But having already passed that point, it is difficult for the companies to say that we are going to donate it to the public domain and to our competitors simultaneously.

Leach: I don't see it as a monopoly by the companies. I see that what they have is of diminishing value, day by day, the more the public sector advances. Ultimately, companies will put their data in the public databases, and they clearly have to make money. I guess more importantly, we in the public sector should be funding the genome projects in advance of the companies. It is time to start bringing it back to the public domain. Unfortunately we duplicate a lot of work in the process.

Gale: Duplication with the private sector is not an issue, because if it is not published it might as well not be done. Zhi-Kang Li was telling me earlier that he heard at a meeting he was at last week that Monsanto have completely sequenced the rice genome and they are going to make it available in a week's time.

Khush: I have also heard the same story.

Li: There was a scientist at the meeting who was called by someone at Monsanto who said that they had decided to release the data to the public domain. Apparently this is from a reliable source.

Mazur: They may be releasing a 4–5× coverage of rice, which won't be enough to identify all genes.

Li: I heard that they were going to release all the data.

Okita: A general comment that I would like to make here is the marked differences in opportunities for research at universities, versus the private sector. Nowadays the forefront of biotechnology research is technology driven. Because much of this new technology is developed by start-up companies, it is usually contracted out to biotech companies and, hence, not normally available to the public sector. Therefore biotech companies have the resources to develop comprehensive databases based on this new technology. I would hope that there would be mechanisms by which the public sector can be made aware of the available databases and given the means to access these databases.

Mazur: We have made our database available on specified terms. Some researchers are unwilling to accept these terms because they don't want commercialization rights to flow to that company. As we condense the number of companies to fewer and fewer, in the end it will not matter that much!

Nevill: Another thought from the industry perspective is that we have to look for areas that are complementary in research. It should never be a 'them and us' scenario. In rice, in many cases there won't be a commercial interest. For example, in Southeast Asia the economy is not right for that sort of information but the data are valuable for public purposes. It is a question of trying to cooperate from an early stage when aims are set.

Regulation of gene expression by small molecules in rice

Shiping Zhang, Lili Chen and Stephen A. Goff[1]

Structural Genomics Group, Novartis Agricultural Discovery Institute, 3115 Merryfield Row, San Diego, CA 92121, USA

Abstract. A system for the regulation of gene expression by small molecules in transgenic rice was developed. This gene switch system consists of two components: (1) a hybrid chemically activated transcription factor, and (2) a synthetic target promoter. The two elements were transformed into rice suspension cells and transgenic plants were regenerated. A luciferase reporter under control of the gene switch system displayed as high as 10 000-fold inducibility following exposure to the small molecule ligand. The dose–response and induction time-course were determined. Regulated luciferase activity in activated plants decreased one day following removal of ligand and could be reactivated multiple times without apparent cosuppression. Analysis of luciferase activity following ligand application to media surrounding the roots suggests that ligand can be absorbed and transported systemically. In contrast, reporter activation was limited to a small area when ligand was applied directly to the leaf surface. The described gene switch system represents an important tool for situations requiring conditional gene expression in a monocot species.

2001 Rice biotechnology: improving yield, stress tolerance and grain quality. Wiley, Chichester (Novartis Foundation Symposium 236) p 85–96

Rice (*Oryza sativa* L.) is the primary food crop for more than a third of the world's population (David 1991). Over the last decade, considerable research effort has been directed at developing rice as a model cereal. Studies on gene position, gene orientation, and gene coding sequences in rice, wheat, and maize have shown that there is significant conservation, both synteny and sequence homology, within the genomes of these important crop plants. In contrast to the conservation of gene order, the rice genome is estimated to be considerably smaller (420–450 million base pairs) than other cereal genomes. For example rice is 1/8th, 1/12th, and 1/38th the size of maize, barley and wheat, respectively (Goff 1999). The smaller genome size of rice, synteny between cereal genomes, and the predicted high gene

---

[1]The chapter was presented at the symposium by Stephen Goff to whom correspondence should be addressed.

density relative to other cereals make rice an attractive model for cereal gene discovery and genomic sequence analysis. An international consortium of publicly-funded researchers has been established to sequence the rice genome within the next several years (*http://rgp.dna.affrc.go.jp/Seqcollab.html*).

Rice transformation technology has progressed rapidly in the past several years (Chen et al 1998b, Christou 1997, Hiei et al 1997). Transformation technology has become a powerful tool for the introduction of foreign genes into rice as well as for the generation of insertional knockout mutations. For example, rice transformation and gene expression technologies have recently led to Taipei 309 rice with enhanced levels of provitamin A (Ye et al 2000), addressing one of the major health concerns in undernourished human populations. In addition, rice transformation via *Agrobacterium* has led to the generation of thousands of lines of rice carrying T-DNA insertional knockouts. Such knockout lines are valuable for the assignment of protein function in genomics studies.

Rice genome sequencing and mutant analysis will yield a large number of interesting genes for further study. Such genes could encode rate-limiting steps in biosynthetic pathways, proteins controlling development, regulators of environmental responses, or predicted coding regions with uncharacterized protein products. A gene switch system will allow the researcher to determine the effect of over-expressing or under-expressing a specific gene of interest. Such a controlled gene expression system will facilitate assignment of a functional role to gene products under study.

Gene switch systems have been developed for a number of different microbes, cell lines, and even intact organisms. For example, derivatives of the mammalian oestrogen receptor have been transferred into yeast to generate an efficient microbial gene switch system (Louvion et al 1993). Likewise, derivatives of the *Drosophila* ecdysone receptor have been transferred into mammalian cell lines and transgenic mice to allow controlled gene expression by invertebrate hormone agonists (Christopherson et al 1992). Schena et al (1991) reported that the mammalian glucocorticoid receptor could regulate gene expression in transiently-expressing tobacco cells. Activation of a CAT reporter gene greater than 150-fold was reported (Schena et al 1991). Additional reports have described plant gene switch systems induced by copper (Mett et al 1993), cytokinin (Faiss et al 1996), and tetracycline (Gatz et al 1992, Weinmann et al 1994). Recent reports have described a glucocorticoid-mediated transcriptional induction system in the model plants tobacco and *Arabidopsis* (Aoyama & Chau 1997), and an oestradiol-inducible gene switch in maize suspension culture cells (Bruce et al 2000). We report here the establishment of oestradiol-controlled gene expression in whole rice plants. The described system will facilitate functional genomic studies in rice and other cereal crops.

Materials and methods

Plasmid constructs.   Plasmid pCIB7613 contains the hygromycin phosphotransferase (*hpt*) gene as a selectable marker for transformation. Plasmid pNADII002 (GAL4-ER-VP16) contains the yeast GAL4 DNA-binding domain (Keegan et al 1986), the mammalian oestrogen receptor ligand binding domain (ER, Greene et al 1986) and the transcriptional activation domain of the HSV VP16 protein (Triezenberg et al 1988). Both *hpt* and GAL4-ER-VP16 were constitutively expressed using the maize ubiquitin promoter. Plasmid pSGCDL1 (GAL4BS Bz1 Luciferase) carries the firefly luciferase reporter gene under control of a minimal maize Bronze1 (Bz1) promoter with 10 upstream synthetic GAL4 binding sites. All constructs use termination signals from the nopaline synthase gene (*nos 3'*, Fig. 1).

Regeneration of inducible transgenic plants.   *Oryza sativa* L. Japonica CV. Taipei 309 were used for the production of the gene switch transgenic rice plants. Callus induction, cell suspension initiation, and maintenance followed protocols previously described by Zhang (1995). Gene transfer was achieved using the Biolistic PDS-1000 system (Bio-Rad. Hercules, CA). Plasmid pCIB7613 was co-transferred with plasmids pNADII002 and pSGCDL1 in a 1:5:5 molar ratio. DNA coating, high-velocity microprojectile delivery of DNA, selection and regeneration of transgenic plants were achieved according to previously published methods (Zhang et al 1998, Chen et al 1998a).

Chemical induction.   Oestradiol (Sigma) was resuspended in 95% ethanol and diluted in water containing 0.01% Triton X-100 immediately before use. The same volume of 95% ethanol was added to the negative control solution without oestradiol. Approximately 10 mg samples of fresh plant tissue were excised and submerged in oestradiol solution, and cultured at 25 °C without light. In whole-plantlet treatments, either intact plantlets were submerged in 1 μM oestradiol solution and cultured at 25 °C in dark for 24 h, or plantlet roots were submerged in $\frac{1}{2}$ MS salts (Murashige & Skoog 1962) liquid medium containing oestradiol followed by incubation at 25 °C with 16 h light cycles.

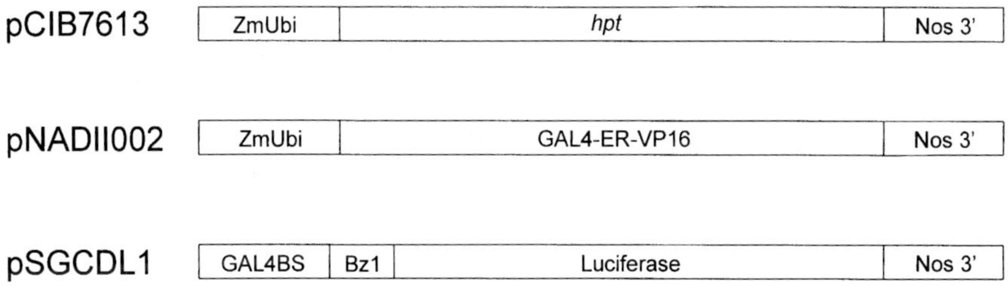

pCIB7613 | ZmUbi | hpt | Nos 3' |

pNADII002 | ZmUbi | GAL4-ER-VP16 | Nos 3' |

pSGCDL1 | GAL4BS | Bz1 | Luciferase | Nos 3' |

FIG. 1.   Maps of plasmid constructs used in this study.

To determine reproducibility of gene switch induction, duplicates (tillers) of plantlets were divided into three groups and intact plantlets were treated with oestradiol solution. Plantlets were washed after the treatment and incubated at 25 °C with 16 h light cycles. Plant leaves were collected at appropriate times, and luciferase activity was determined. After luciferase activity decreased to background, plantlets were reinduced with oestradiol.

To determine if ligand acts systemically, trimmed leaf tip ends of inducible plants were continuously exposed to 1, 10 or 50 µM oestradiol solutions *in vivo*. Freshly prepared solutions were exchanged daily. Samples from leaf tips, exposed to oestradiol solution, 5 and 10 cm from the exposed leaf tips were collected after one and four day treatments.

Luciferase assays.   Treated and control plant tissues were frozen in liquid nitrogen and ground to a fine powder then resuspended in 200 µl cell lysis reagent (Promega Cat. No. E153A, Madison,WI). Cell debris was pelleted for 10 min in an Eppendorf microfuge. Protein content of the resulting supernatant was determined using the Bradford protein assay kit (Bio-Rad Cat. No. 500-0006. Hercules, CA) according to the manufacturer's instructions. 30 µl of tissue extract was used to determine luciferase activity in a Berthod luminometer (Lumat LB 9507, EG&G Berthod, Gaithersburg, MD) with 100 µl luciferase assay reagent (Promega Cat. No. E1501, Madison, WI). Luciferase data are expressed as relative light units (RLU) per microgram soluble protein per 10 seconds.

Detection of luciferase luminescence.   Following oestradiol treatment, intact plantlets were submerged in 2 mM Luciferin (Biosynth AG Cat. No. L-8220, Naperville, IL) with 0.01% Triton X-100 solution for 10 min. Luciferase luminescence photographs were taken using a Multilmage™ light cabinet (Alpha Innotech Corp, San Leandro, CA) at 10 min exposure times. After photographs were taken, plantlets were returned to growth solution without oestradiol, treated with luciferin, and photographed at 12 h intervals.

Results and discussion

A total of 35 independent hygromycin-resistant transgenic rice plant lines were regenerated. Five lines were found to display oestradiol-inducible luciferase activity. Selfed R1 progeny plants of a single transgenic line were used for all reported experiments.

Dose–response of gene switch activation

To determine the range of induction with varying concentration of ligand, plant leaves were exposed to 0, 0.001, 0.01, 0.1, 1, 10 or 100 µM oestradiol for 24 h, and

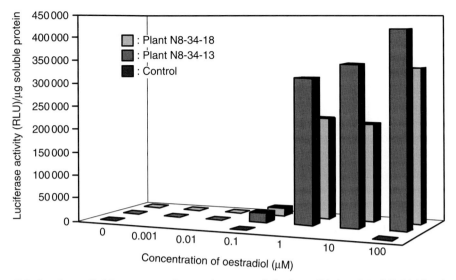

FIG. 2. Oestradiol dose response in two rice gene switch plants. R1 rice plant (N8-34-13 and N8-34-18) leaf samples were cultured *in vitro* in oestradiol solution for 24 h. Each value is the mean of three independent replicates.

luciferase reporter activity was determined (Fig. 2). Luciferase reporter activity remained near background in plant tissues treated with 10 nM oestradiol or less. At 100 nM oestradiol exposure, a significant increase in luciferase activity was observed. At 1 μM oestradiol, luciferase reporter activity was more than 20 000-fold higher than in plants untreated with ligand. Increasing oestradiol concentrations to 100 μM did not significantly enhance reporter activity, suggesting the gene switch was maximally induced by 1 μM oestradiol. These dose responses are consistent with those observed in other systems containing ligand-binding domains from mammalian steroid or nuclear receptors. An oestradiol concentration of 1 μM was used as the standard induction dose in the following experiments.

Time course of gene switch activation

To determine how rapidly the gene switch responds to the addition of ligand, we examined the time-course of luciferase activity following exposure of plant leaves to ligand (Fig. 3). Luciferase activity increased approximately 30-fold after a 3 h oestradiol treatment period. Luciferase activity increased 300-fold following 6 h of treatment with ligand. Maximum induction of approximately 10 000-fold was observed at 24 h of oestradiol treatment (Fig. 3). This relatively rapid time course of induction is consistent with the reported molecular mode of action of steroid

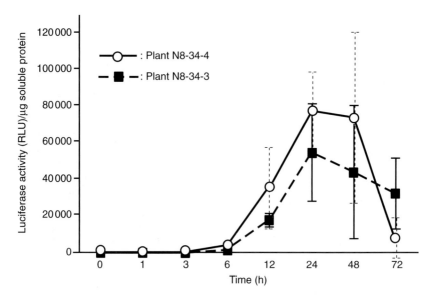

FIG. 3.   Time course of luciferase induction in rice gene switch plants. R1 rice plant (N8-34-3 and N8-34-4) leaf samples were cultured *in vitro* in 1 µM oestradiol solution and collected for luciferase activity assay over time. Each value is the mean of three replicates, with error bars representing ± standard deviation.

receptors. It is believed that the receptor exists in an unfolded state bound to molecular chaperonins in the absence of ligand, and proper folding and activation of the receptor is facilitated by the presence of ligand (Picard 1993). Further periods of plant leaf inoculation in oestradiol solution resulted in a decrease of luciferase activity. This may be due to plant cell death during the longer anaerobic condition in the solution.

Repeated gene switch activation over time

To examine the ability of the gene switch in a single plant or tissue to be repeatedly induced by ligand, we used transient exposure and removal of ligand. Duplicate tillers of intact plantlets were treated with oestradiol solution, and luciferase activity was followed after exposure and removal of ligand and re-exposure to ligand (Fig. 4). Luciferase activity reached a maximum approximately 1 day after exposure and removal of oestradiol. Reporter activity then dramatically decreased in the following 3 days, returning to the level of an untreated plantlet in approximately 20 days. Luciferase activity could be enhanced to the maximal level by additional oestradiol treatment periods following the initial removal. As

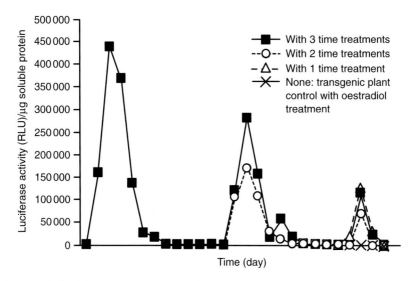

FIG. 4.   Reinducibility in rice gene switch plants. R1 rice plants N8-34-60 and N8-34-72 were used. Intact plantlets were cultured for 24 h *in vivo* in 1 μM oestradiol solution starting at day 0 (1st treatment), day 21(2nd treatment) and day 39 (3rd treatment). Leaf samples were collected at various times after induction and luciferase activity was assayed. Each value is the mean of 6 replicates.

many as three gene switch activations could be applied without a significant change in the resulting reporter activation. These observations suggest that the luciferase reporter in this transgenic line is not subject to gene silencing following the initial induction regime. The data in this and other experiments also indicate that the peak induction level decreases with plant age. This may be because young plant leaves can be easily passed through by ligands and thus more cells received the chemicals.

Detection of ligand uptake and movement in plant leaves

To determine the ability of oestradiol ligand to move from the tips of exposed leaves toward the leaf base, we determined luciferase reporter activities in regions of leaves directly exposed to ligand and adjacent regions unexposed to ligand. Leaf tips were clipped and the exposed surface was treated with 1 μM oestradiol for 1–4 days. Leaf samples at the site of treatment as well as samples 5 and 10 cm toward the base of the leaf were harvested and assayed for luciferase activity. A 22 000–80 000-fold gene switch activation was found after one and four days of 1 μM oestradiol treatment of the ligand-exposed leaf tips (Fig. 5). In contrast to this high level of activation at the leaf tips, there were no significant increases in reporter activity in leaf samples collected 5 and 10 cm from oestradiol-treated tips. With higher ligand concentrations of 10 and 50 μM during the treatment, a 10–200-fold range increase

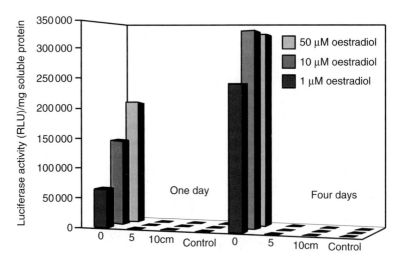

FIG. 5. Ligand movement away from leaf tips in gene switch plants. Leaf tips were continuously exposed to 1, 10 or 50 μM oestradiol solution (replaced daily), and samples various distances from the exposed tips were assayed for luciferase activity. Each value is the mean of three replicates.

in luciferase activity in the 5 and 10 cm basal regions was observed. Although this increase is significant, it was considerably below the 40 000-fold induction observed in exposed leaf-tip regions. These observations suggest that there is little movement of oestradiol ligand from clipped leaf tips toward the base of the leaf.

Detection of luciferase bioluminescence in intact plants

Intact plantlets treated with oestradiol by immersion of the entire plantlet into ligand solution displayed high luminescence in leaf tissue, but not in roots or meristematic tissues (Fig. 6). These results are consistent with reporter activities obtained from *in vitro* treatment of various plant tissues.

All plant tissues showed luminescence light when plantlets were exposed to oestradiol solution via root immersion. These results suggest that ligand can be absorbed by roots and transported systemically. However, plants treated in this fashion displayed a lower luminescence in leaves when compared to leaf tissue treated directly with ligand. This observation suggests that systemic movement of ligand is limited or degradation of ligand is occurring following root absorption. Luciferase activity was detectable for approximately 72 h after oestradiol removal (Fig. 7).

A **B**

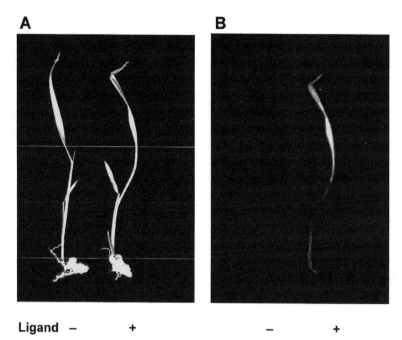

Ligand – + – +

FIG. 6. Chemical regulation in intact rice plants. Intact plantlets were *in vivo* cultured in 1 μM oestradiol solution for 24 h, then immersed in 2 mM luciferin for 10 min. Luminescence photographs were taken in an Alpha Innotech Corp. MultiImage light cabinet with (A) and without internal light (B, 10 min exposure).

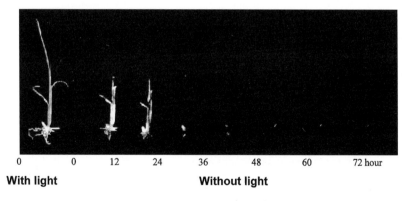

0   0   12   24   36   48   60   72 hour

With light   **Without light**

FIG. 7. Duration of luciferase activity in induced plants. Plant roots were exposed to 1 μM oestradiol solution for 24 h, and light detection photographs were taken at 12 h intervals after 10 min luciferin treatments. Luciferase luminescence pictures were taken in an Alpha Innotech Corp. MultiImage light cabinet with and without light (10 min exposures).

In conclusion, we have established a gene switch system for intact rice plants. This system provides for conditional expression of transgenes in intact rice plants. This gene switch system will be highly useful for ongoing and future functional genomic studies.

Acknowledgements

The authors thank Novartis Agricultural Biotechnology Research Institute for providing the plasmids pSGCDL1 and pCIB7613.

References

Aoyama T, Chau NH 1997 A glucocorticoid-mediated transcriptional induction system in transgenic plants. Plant J 11:605–612

Bruce W, Folkerts O, Garnaat C, Crasta O, Roth B, Bowen B 2000 Expression profiling of the maize flavonoid pathway genes controlled by estradiol-inducible transcription factors CRC and P. Plant Cell 12:65–79

Chen L, Marmey P, Taylor N, Brizard JP et al 1998a Expression and inheritance of multiple transgenes in rice plants. Nat Biotechnol 16:1060–1064

Chen L, Zhang S, Beachy RN, Fauquet CM 1998b A protocol for consistent, large-scale production of fertile transgenic rice plants. Plant Cell Rep 18:25–31

Christopherson KS, Mark MR, Bajaj V, Godowski PJ 1992 Ecdysteroid-dependent regulation of genes in mammalian cells by a *Drosophila* ecdysone receptor and chimeric transactivators. Proc Natl Acad Sci USA 89:6314–6318

Christou P 1997 Rice transformation: bombardment. Plant Mol Biol 35:197–203

David C 1991 The world rice economy: challenges ahead. In: Khush GS, Toenniessen GH (eds) Rice Biotechnology. CAB International, Wallingford, p 1–18

Faiss M, Strnad M, Redig P et al 1996 Chemically induced expression of the *roIC*-encoded β-glucosidase in transgenic tobacco plants and analysis of cytokinin metabolism: roIC does not hydrolyze endogenous cytokinin glucosides *in planta*. Plant J 10:33–46

Gatz C, Frohberg C, Wendenburg R 1992 Stringent repression and homogeneous de-repression by tetracycline of a modified CaMV 35S promoter in intact transgenic tobacco plants. Plant J 2:397–404

Goff SA 1999 Rice as a model for cereal genomics. Curr Opin Plant Biol 2:86–89

Greene GL, Gilna P, Waterfield M, Baker A, Hort Y, Shine J 1986 Sequence and expression of human estrogen receptor complementary DNA. Science 231:1150–1154

Hiei Y, Komari T, Kubo T 1997 Transformation of rice mediated by *Agrobacterium tumefaciens*. Plant Mol Biol 35:205–218

Keegan L, Gill G, Ptashne M 1986 Separation of DNA binding from the transcription-activating function of eukaryotic regulatory protein. Science 231:699–704

Louvion JF, Havaux-Copf B, Picard D 1993 Fusion of GAL4-VP16 to a steroid-binding domain provides a tool for gratuitous induction of galactose-responsive genes in yeast. Gene 131:129–134

Mett VL, Lochhead LP, Reynold PH 1993 Copper-controllable gene expression system for whole plants. Proc Natl Acad Sci USA 90:4567–4571

Murashige T, Skoog F 1962 A revised medium for rapid growth and bioassays with tobacco tissue cultures. Physiol Plant 15:473–497

Picard D 1993 Steroid-binding domains for regulating the functions of heterologous proteins *in cis*. Trends Cell Biol 3:278–280

Schena M, Lloyd AM, Davis RW 1991 A steroid-inducible gene expression system for plant cells. Proc Natl Acad Sci USA 88:10421–10425

Triezenberg S J, Kingsbury RC, McKnight SL 1988 Functional dissection of VP16, the transactivator of herpes simplex virus immediate early gene expression. Genes Dev 2:718–729

Weinmann P, Gossen M, Hillen W, Bujard H, Gatz C 1994 A chimeric transactivator allows tetracycline-responsive gene expression in whole plants. Plant J 5:559–569

Ye X, Al-Babili S, Klöti A et al 2000 Engineering the provitamin A (beta-carotene) biosynthetic pathway into (carotenoid-free) rice endosperm. Science 287:303–305

Zhang S, Song WY, Chen L et al 1998 Transgenic elite Indica rice varieties, resistant to *Xanthomonas oryzae* pv. *Oryzae*. Mol Breed 4:551–558

Zhang S 1995 Efficient plant regeneration from protoplasts of four true Indica (group 1) rice varieties and advanced line. Plant Cell Rep 15:68–71

DISCUSSION

Mazur: Have you been able to combine chemical inducibility and tissue specificity in a single construct?

Goff: We haven't done this yet, but you can see how it would fall out of this work fairly easily. We would need to express the receptors under a tissue-specific promoter, and then these should be the only tissues that respond.

Gale: What are your main targets? Why do you want to do this?

Goff: A field gene switch has applications like hybridization, or the overproduction of an otherwise detrimental protein in a constitutive state. For research purposes, overproduction or generating some inhibitory protein domain would help assign function. If a set of genes are of particular interest, it may be useful to know the phenotypic consequences of overexpressing them.

Gale: It sounds great, but I don't know how you will get it into the public domain without it having the same stigma that the terminator gene technology has. This is a gene-use restriction technology that you are putting together here.

Goff: It is not a terminator technology. The transgenes are only active when you want them to be. If there is a detrimental effect of expressing the transgene, then you probably won't want them on all the time.

Mazur: There is the possibility for creating a terminator-like effect, but the wider utility is for controlling something detrimental to the plant that has value as a product.

Salmeron: In most situations it would be up to the farmer whether they wanted to activate this.

Gale: I understand that, but it is still a gene-use restriction technology.

Mazur: One example of its use would be if you wanted to make a polymer in a plant, and wanted to turn on its synthesis only in the final grain production.

Leach: You induced it by feeding the hormone through the roots. Can you induce it by spraying?

Goff: We usually induce it by painting it on, but spraying would work. We think it is systemic, but this will not be the case for every compound.

Dong: There was a report in *Plant Journal* recently (Kang et al 1999) showing that the VP16-GAL4 gene is toxic to *Arabidopsis*, even in the absence of dexamethasone.

Goff: Similar experiments were done with different constructs in *Arabidopsis*, and the lines appear unstable, unlike rice. It's unclear why.

Parker: Do you find that the oestradiol system is more robust in *Arabidopsis*? I understand that oestradiol does not travel systemically, whereas dexamethasone does. Is that correct?

Goff: There are currently not enough data in *Arabidopsis*. I think Jeff Dangl's lab may have more data on this system.

Reference

Kang HG, Fang Y, Singh KB 1999 A glucocorticoid-inducible transcription system causes severe growth defects in Arabidopsis and induces defence-related genes. Plant J 20:127–133

General discussion II

Ku: Mike Gale, how would you go about using synteny to clone a gene?

Gale: If you work in wheat, the attraction of the synteny approach is that you can identify a gene in a small genome and then go back and pull out the wheat equivalent. There are several examples of this sort of work going on. If the approach works, particularly when we have the entire rice sequence, this is a means to isolate your favourite genes from any crop—even those in which we don't have any genomic tools.

Ku: We have been working on C4 photosynthesis for a long time. We know quite a bit about the C4 biochemistry, but we know nothing about the genetics behind the Kranz anatomy. How can we go about looking for these subtle evolutionary changes, such as the switch from C3 to C4 photosynthesis?

Gale: I really don't know enough about the system to comment on how you would best look for this. However, if you have a gene in any one species, you can certainly go straight in with a Southern analysis to demonstrate whether a homologue is present in another species.

Leung: In a recent photosynthesis meeting Bill Taylor from CSIRO mentioned the genomic approach for looking at the C3/C4 pathways. Apparently, maize and rice will useful for looking at this difference.

Bennett: One way forward would be to screen the germplasm of the C4 crop in search of mutants or variants that lacked Kranz anatomy and yet could be crossed with the cultivars that possessed this architecture. This would allow mapping of the genes controlling formation of Kranz anatomy. Using synteny, you could isolate the corresponding genes from rice.

Ku: In the natural population there are no Kranz anatomy-deficient mutants. Perhaps it would be a lethal phenotype in terms of survival under ambient conditions. We tried to collaborate with Virginia Walbot who has a maize transposon tagging population. Perhaps we will be able to find a Kranz anatomy-deficient mutant this way.

Bennett: I am still fascinated by the sequences between the genes, which seem to be responsible for the total size of the genome. What is the total recombination length of a diploid *Triticum*?

Gale: The recombination lengths are pretty much around 100 map units per chromosome, no matter how much DNA there is in the chromosome. If you say that all recombination takes place in genes, this will account for that. There is no

97

relationship between total DNA and recombination, or so it seems. Wheat chromosomes are literally 30 times as big as rice chromosomes, and the answer is the same for both.

Bennett: These elements may be kept in check by methylation. Is it possible to envision that a period without methylation might have led to the expansion of the genome?

Gale: You can date these elements by mutation and base pair changes. Jeff Bennetzen's lab has done that, and the answer is remarkably that most of the transposition took place between 3 and 6 million years ago (SanMiguel et al 1998). There was apparently a period of massive expansion of the maize genome. Maize has been around for the best part of 21 million years, yet all the genome expansion took place in a small period relatively recently.

Leung: How can we fast track application of expressed sequence tags (ESTs) to look at allelic diversity? Mike Gale thinks that microsatellites are the best way of characterizing germplasm. In my mind, isn't a more direct way to take putative functional genes and use them to look at allelic diversity? In the days of isozyme analysis, the advantage was in using neutral markers to look at evolution. If we look at agronomic traits, isn't EST profiling a more direct way of predicting function?

Mazur: I think what you are asking for is a single nucleotide polymorphism (SNP) analysis of all the different varieties, looking for specific allelic differences.

Gale: The reason why microsatellites are attractive is that they are a very polymorphic system, and it is a multiple allelic system if it is looked at on a gel. SNPs aren't; they are essentially alternative alleles. In the old days we spent a lot of time with cDNAs, trying to make sequence tag site (STS) markers and develop polymorphism in that way. You can look at hundreds of varieties and never see any polymorphism at all. Of course, if you can instantly get into single-nucleotide changes, it may be a different game. For crops like wheat and rice today, microsatellites are really the only feasible technique. John Bennett, you have spent a lot of time looking at STS from cDNA clones. Would you recommend that as a way to find polymorphism?

Bennett: I think I would prefer them to microsatellites, which can be 5–20 cMs away from a gene of interest. I think microsatellites are good for germplasm analysis, but not for allele mining because they don't relate to a gene that you are interested in. The way we have been using STS is to see how far out from japonicas we can go with japonica 5′ and 3′ primers. In fact, we can go all the way out into the wild rices. The primers that you would design from the 5′ and 3′ end of a nipponbara gene will amplify a single gene from all the AA genome rices and most of the non-AA genome rices. They start to fall down with some of the non-AA genome species, the same ones in each case. If you go to the wild relatives of rice, non-*Oryzae* species, they don't amplify, and they don't amplify for the other cereals. This is just taking the 5' and 3' regions, using PCR to clone alleles. At the moment I don't think we have a simple way of allele mining. If you want the entire gene and you want your primers to be upstream of the promoter and downstream of the terminator, how far out can you use PCR and mine those alleles? I would divide allele mining into two steps: actually getting the alleles in your hand and then finding out whether they are of any interest. There are so many different sorts of alleles, I know of no method of getting at them all conveniently.

Ku: Quite a few years ago Toryama, in Japan, tried to fuse protoplasts isolated from a C3 plant (rice) and a C4 plant (*Echinochloa crysgalli*). They ended up with quite a few somatic hybrids, but most of them did not show good chromosome pairing. Eventually they all died.

Gale: I don't know of any examples of artificial species that have been made in rice.

Khush: There is a cross between *Oryza sativa* and *Portresia coarctata*. The hybrids are viable but there is no fertility or chromosome pairing.

Gale: There doesn't have to be pairing between the chromosomes of the two species. You make the F1, which will normally be sterile, and then double the chromosomes. You can do this in a number of ways but the use of colchicine is a common method. This is how *Triticale* was formed. Hexaploid *Triticale* ($2n = 6 \times = 42$) derives from a cross between tetraploid wheat ($2n = 4 \times = 28$) and diploid rye ($2n = 2 \times = 14$). The sterile 21-chromosome hybrid was doubled to 42 giving an amphiloid with a normal diploid meiosis.

Khush: There are many examples like that where when the chromosomes of interspecific hybrids are doubled, the fertility is not restored all the time. *Triticale* is a rare example where there is good fertility and where, by selection, it has been possible to develop an agronomically useful plant. But there are many examples where even the amphiploids are completely sterile.

Li: Interactions between genes from different genomes, even from different subspecies of rice like japonica and indica, may cause hybrid breakdown (sterility in F_2 or advanced progenies), which is a major mechanism to block the gene flow. Wheat is rare in that you can simply double the chromosome number and get a fertile progeny, leading to a new species. In the majority of cases, when the genetic divergence reaches a certain level, you won't easily be able to put different genomes together.

Reference

SanMiguel P, Gaut BS, Tikhonov A, Nakajima Y, Bennetzen JL 1998 The paleontology of intergene retrotransposons of maize. Nat Genet 20:43–45

Introduction of genes encoding C4 photosynthesis enzymes into rice plants: physiological consequences

Maurice S. B. Ku, Dongha Cho*, Xia Li†, De-Mao Jiao†, Manuel Pinto‡, Mitsue Miyao¶ and Makoto Matsuoka§

*School of Biological Sciences, Washington State University, Pullman, WA 99164-4236, USA, *Division of Applied Plant Sciences, College of Agricultural and Life Sciences, Kangwon National University, Kangwon, Korea, †Institute of Agrobiological Genetics and Physiology, Jiangsu Academy of Agricultural Sciences, Nanjing 210014, China, ‡Facultad de Ciencias Agrarias y Forestales, Universidad de Chile, Santa Rosa 11315, Santiago, Chile, ¶National Institute of Agrobiological Resources, Tsukuba 305-8602, Japan, and §BioScience Center, Nagoya University, Chikusa, Nagoya 464-8601, Japan*

Abstract. Transgenic rice plants expressing the maize phospho*eno*/pyruvate carboxylase (PEPC) and pyruvate, orthophosphate dikinase (PPDK) exhibit a higher photosynthetic capacity (up to 35%) than untransformed plants. The increased photosynthetic capacity in these plants is mainly associated with an enhanced stomatal conductance and a higher internal CO_2 concentration. Plants simultaneously expressing high levels of both enzymes also have a higher photosynthetic capacity. The results suggest that both PEPC and PPDK play a key role in organic acid metabolism in the guard cells to regulate stomatal opening. Under photoinhibitory and photooxidative conditions, PEPC transgenic rice plants are capable of maintaining a higher photosynthetic rate, a higher photosynthetic quantum yield by PSII and a higher capacity to dissipate excess energy photochemically and non-photochemically than untransformed plants. Preliminary data from field trials show that relative to untransformed plants, the grain yield is about 10–20% higher in selected PEPC and 30–35% higher in PPDK transgenic rice plants, due to increased tiller number. Taken together, these results suggest that introduction of C4 photosynthesis enzymes into rice has a good potential to enhance its tolerance to stress, photosynthetic capacity and yield.

2001 Rice biotechnology: improving yield, stress tolerance and grain quality. Wiley, Chichester (Novartis Foundation Symposium 236) p 100–116

From a photosynthetic point of view, crop yield can be further improved by increasing the photosynthetic capacity in source leaves and/or by increasing partitioning of photoassimilate from the source leaves to organs of economic interest. In a comparative study with eight wheat cultivars released by CIMMYT (Centro Internacional de Mejoramiento de Maiz y Trigo) between 1962 and 1988

(Fischer et al 1998), it was discovered that the progressive increases in grain yield in the new strains (27%) are highly correlated with enhanced stomatal conductance to CO_2 diffusion (63%). The enhanced stomatal conductance results in higher photosynthetic rates on a leaf area basis (23%). Thus the breeders had screened for high-yielding strains by selecting unknowingly for higher photosynthetic capacity in the source leaves. Another benefit of enhanced stomatal conductance is depression of leaf canopy temperature (0.6 °C), which reduces heat stress. An earlier analysis by Gifford et al (1984) suggested that an increase in partitioning of photoassimilate to harvested organs has been of primary importance for the improvement of genetic yield potential of crops. The progressive increases in yield for crops such as wheat, barley, oat and soybean in the past century have not been associated with increases in crop biomass, but with increased harvest index (distribution of biomass between economic yield and the rest of the plant). The increases in grain yield in the eight wheat cultivars released by CIMMYT are also related to improved harvest index. Therefore, a higher photosynthetic capacity in the source leaves coupled with increased partitioning of photoassimilate to harvested organs holds great promise for improving crop productivity. However, the biochemical bases for these important agonomic traits remain unknown.

The ultimate goal of our research is to improve the photosynthetic efficiency of crops that assimilate atmospheric CO_2 via the C3 pathway of photosynthesis by introducing some of the features associated with C4 photosynthesis. C4 plants have many desirable agronomic traits including high photosynthetic capacity and high mineral use efficiency, especially under high light, high temperature and drought conditions, due to the CO_2-concentrating mechanism by the additional C4 pathway of photosynthesis (Hatch 1987, Ku et al 1996). On the other hand, plants that utilize the conventional C3 pathway for carbon fixation, including many of our agronomically important species such as rice and wheat, suffer from O_2 inhibition of photosynthesis and the associated photorespiration, and exhibit a lower photosynthetic efficiency under these conditions. Photosynthetically, these plants are underachievers because, on the one hand, they assimilate atmospheric CO_2 into sugars but, on the other hand, part of the potential for sugar production is being lost by respiration in the light, releasing CO_2 into the atmosphere. This is due to the dual function of the key photosynthetic enzyme, ribulose 1,5-bisphosphate carboxylase/oxygenase (Rubisco). High CO_2 favours the carboxylase reaction and thus net photosynthesis; whereas high O_2 promotes the oxygenase reaction leading to photorespiration. When plants first evolved, photorespiration was not a problem because the atmosphere then was high in CO_2 and low in O_2. As a byproduct of photosynthesis, O_2 accumulated in the atmosphere and reached the current level (21%) hundreds of million years ago. On the other hand, the atmospheric CO_2 levels have decreased throughout geologic time (see Ehleringer et al 1991). Current atmospheric CO_2 levels (0.036%) limit photosynthesis in C3 plants. Furthermore, photorespiration reduces net carbon gain and productivity of C3 plants by as much as 40%. This renders C3 plants less competitive in certain environments — high light, high temperature and drought conditions. With some modifications in leaf anatomy, some tropical species (e.g. maize and sugarcane) have evolved a biochemical 'CO_2 pump', the C4 pathway of photosynthesis, to concentrate the atmospheric CO_2 in the inner bundle sheath cells where Rubisco is located and overcome photorespiration. Thus C4 photosynthesis is not significantly limited by the current levels of atmospheric CO_2.

The conventional cross-hybridization approach has been employed to transfer C4 traits to C3 plants (see Brown & Bouton 1993). However, epistatic interaction between the alleles suppresses the expression of C4 traits in the progeny, and genes for Kranz leaf anatomy and biochemistry of C4 photosynthesis are not closely linked. A large segregation population will be required to find a progeny expressing both features. Most importantly, no closely related C3 and C4 crops can be hybridized. With the advancement in molecular biology, recombinant DNA technology has been used to introduce some of the C4 features into C3 crops. Several attempts have been made in the past to express the genes encoding the enzymes of the C4 pathway in C3 plants in an effort to tune up their photosynthetic metabolism (Hudspeth et al 1991, Kogami et al 1994, Gehlen et al 1996, Gallardo et al 1995, Ishimaru et al 1998). However, limited physiological consequences were observed in these transgenic plants, presumably due to low levels of expression of these genes.

In engineering the CO_2-concentrating mechanism of C_4 photosynthesis, there are two important components to be considered: the biochemical pathway (enzymes) and the specialized leaf structure. The coordination of two specialized leaf cells in C4 leaves, namely mesophyll and bundle sheath cells (together termed Kranz leaf anatomy), is important for pathway function. The enzymes and their corresponding genes involved in the C4 pathway of photosynthesis have been characterized. However, very little is known about the molecular mechanisms controlling the differentiation of Kranz anatomy in C4 plants. Therefore, our first goal was to engineer the key enzymes involved in C4 photosynthesis in rice without Kranz leaf anatomy. At first thought, one may argue that rice plants thus engineered may not be very efficient in concentrating CO_2 in the leaf, as Rubisco is located in the chloroplasts of the inner bundle sheath cells in C4 leaves. The cell wall of these well-differentiated inner cells has special constituents to prevent CO_2 from leaking out of the leaf. However, in nature, a primitive aquatic plant, *Hydrilla verticillata*, is known to be able to use a simplified version of the C4 pathway to concentrate CO_2 and eliminate the wasteful photorespiration process (Reiskind et al 1997, Magnin et al 1997). When it is grown under low CO_2 conditions,

H. verticillata shifts from C3 to C4 photosynthesis and assimilates atmospheric CO_2 via the C4 pathway without Kranz leaf anatomy (Bowes & Salvucci 1989). Inorganic carbon is first assimilated into the C4 acid malate in the cytoplasm via phospho*enol*/pyruvate carboxylase (PEPC), and subsequently malate serves as a donor of CO_2 to Rubisco in the chloroplast by the decarboxylating enzyme, NADP-malic enzyme (NADP-ME). This primitive type C4 photosynthesis is sufficient to concentrate CO_2 in the chloroplast and overcome photorespiration (Reiskind et al 1997). It is possible that this archetypal version of C4 photosynthesis, which does not depend on Kranz compartmentation, can be engineered to function in terrestrial C3 plants.

We have independently introduced three key C4 photosynthesis genes from maize into rice: PEPC, pyruvate, orthophosphate dikinase (PPDK) and NADP-ME (Agarie et al 1998, Ku et al 1999). First, it is hoped that by introducing some of the key enzymes of C4 photosynthesis into C3 plants with proper intercellular compartmentation a limited C4 acid metabolism may be installed for fixing atmospheric CO_2 directly via this pathway and partially concentrating CO_2 in the chloroplast, as exhibited by *H. verticillata*. Second, enhanced expression of the enzymes of C4 photosynthesis in C3 plants may increase carbon and nitrogen metabolism in certain tissues of C3 plants. All enzymes involved in C4 photosynthesis are found in leaves of C3 plants. As in C4 plants, they catalyze similar reactions in C3 leaves, but do not contribute significantly to overall CO_2 assimilation. Although they are low in activity in leaves of C3 plants, some of them are found at high levels in reproductive tissues. For example, the cytosolic isoform of PPDK occurs at high levels in seeds of both the C3 plant wheat (Aoyagi & Bassham 1984a, b, Aoyagi & Chua 1988, Blanke & Lenz 1989) and the C4 plant maize (Imaizumi et al 1997). It may play an important role in carbon and nitrogen metabolism or energy supply (e.g. release of ATP from PEP catalyzed by PPDK) in reproductive tissues, and enhanced expression of the enzyme may boost seed development and grain productivity. Third, enzymes involved in C4 photosynthesis, albeit low in C3 plants, may also play important roles in plant defence responses to biotic and abiotic stress. Metabolic alterations in response to stress allow plants to adapt to adverse conditions. For example, up-regulation of NADP-ME by wounding, low oxygen, low temperature, salinity, and ultraviolet light has been reported in both the C3 plants rice (Fushimi et al 1994) and bean (Walter et al 1994, Schaaf et al 1995, Pinto et al 1999) and the C4 plant maize (Drincovich et al 1998). It is postulated that the reductant (NADPH) released from decarboxylation of malate by NADP-ME may be required for the increased synthesis of secondary metabolites for defence purpose. Furthermore, increased expression of PPDK in C3 chloroplasts may enhance synthesis of aromatic amino acids such as phenylalanine via the shikimic pathway (Hermann 1995), which serves as the substrate for biosynthesis of secondary metabolites (e.g.

phenylpropanoids) involved in plant defence mechanism (Douglas 1996). The biosynthesis of phenylpropanoids requires an efficient flow of carbon into phenylalanine. Thus, increased expression of some C4 photosynthesis enzymes in C3 plants could confer enhanced tolerance under stress conditions.

Rice transformation and expression of transgenes

Using the *Agrobacterium*-mediated transformation system, we have transformed two Japonica rice cultivars, Kitaake and Nipponbare, with three maize C4 photosynthesis genes, as mentioned above (Agarie et al 1998, Ku et al 1999). Transgenic plants derived from Kitaake were used for further characterization since Kitaake has a shorter life span and flowers more readily under our growth conditions. The level of expression varies considerably among the transgenic plants derived from the same transformation. In part, this could be due to site of insertion of the transgene in the rice genome, but it is also dependent on gene construct used for transformation and gene copy. In comparison, intact maize genes with their own promoter and terminator sequences and introns and exons tend to give higher levels of expression, as compared with cDNAs (Matsuoka et al 2000). In some transgenic plants harbouring the intact maize PEPC gene, the maize enzyme accounts for up to 18% of the total leaf soluble protein (Ku et al 1999). Also, the level of expression is related to gene copy number and locus number of transgene insertion. Genetic studies show that the maize genes are stably inherited in a Mendelian manner, with the genes being inserted at one or two loci into the rice genome. Immunolocalization studies show that the maize PEPC is expressed in the cytosol whereas PPDK and NADP-ME are expressed in the chloroplast, as one would expect. The maize enzymes remain active in the transgenic rice plants, and the activity is highly correlated with the enzyme protein amount (Ku et al 1999). These results suggest that the regulatory mechanisms for maintaining the activity of the maize C4 enzymes are present in rice. In terms of organ specificity of expression, both leaf and leaf sheath of transgenic rice plants express high levels of these enzymes. A substantial amount of PEPC is also detected in the palea and lemma tissues of seed in PEPC transgenic plants. In contrast, these enzymes are predominantly expressed in leaves of maize. The results suggest that the regulatory mechanisms responsible for organ-specific expression of these genes in maize is not present in C3 rice.

Except for few plants, most of the transgenic rice plants exhibit a normal phenotype and retain high fertility (85–90% as versus 90% in the untransformed plants). However, NADP-ME transgenic plants have reduced height (10 cm), and the time required to flower is shortened by 4 days for PPDK transgenic plants but delayed 4–6 days in PEPC transgenic plants. Another interesting observation is that leaves of NADP-ME transgenic plants stay green even after the seeds have reached maturity. The reason for this is not clear, but may be related to the production of extra reductant via NADP-ME for chlorophyll synthesis. Maintenance of photosynthetically active leaves during grain filling will further contribute to yield potential.

Photosynthetic performance of transgenic rice plants expressing maize C4 photosynthesis enzymes

PEPC transgenic rice plants

On a leaf area basis, the photosynthetic rates of the primary PEPC transgenic plants, measured under ambient conditions, are comparable or higher than those of untransformed plants (Ku et al 1999). In addition, O_2 inhibition of photosynthesis decreases progressively with increasing level of PEPC activity among the transgenic plants. Our preliminary labelling experiment with $^{14}CO_2$ shows only a small increase (4%) in atmospheric CO_2 being directly fixed by PEPC in these plants. The supply of PEP, the substrate for PEPC, may be limited in C3 leaves. Thus, the biochemical and physiological bases of these alterations in photosynthetic trait remain unclear. Using the segregation populations from four primary transgenic lines which exhibit high levels of the maize PEPC, we have shown that the photosynthetic rates of flag leaves in most PEPC transgenic rice plants are comparable or up to 30% higher than those of untransformed plants (Ku et al 2000). Photosynthetic rate begins to decrease as the level of expression reaches very high, as one would expect. Indeed, transgenic plants with extremely high levels of PEPC have lower chlorophyll contents. Analysis of the relationship between photosynthetic rate and stomatal conductance among these plants shows a good positive correlation between the two parameters. Furthermore, stomatal conductance is highly correlated with intercellular CO_2 concentration, and the intercellular CO_2 concentration in some transgenic plants is as high as $275\,\mu l\,l^{-1}$, in comparison to $235\,\mu l\,l^{-1}$ in untransformed plants. Therefore, part of the higher photosynthetic capacity of the transgenic plants may be due to the ability of the plants to maintain a higher internal CO_2 in the leaf due to increased stomatal opening. The immediate benefit of a higher intercellular CO_2 is elevated net carbon fixation due to more CO_2 and suppression of Rubisco oxygenase and the associated photorespiration. Consistently, we also observed an upward shift in optimal temperature for photosynthesis by the transgenic plants from 26 to 28–32 °C (U. Ranade & M. S. B. Ku, unpublished data), presumably due to reduced photorespiration.

Consistent with the suggestion that PEPC transgenic rice plants have a higher stomatal conductance than untransformed wild-type plants, the $\delta^{13}C$ values for the transgenic plants are 1.5–2.5‰ more negative than that of untransformed plants and the value increases with increasing PEPC activity among the transgenic plants (Ku et al 2000). An increased stomatal conductance would allow more CO_2 to diffuse into the leaf and thus more ^{13}C being discriminated during photosynthesis (Winter et al 1982). However, the possibility that the lower ^{13}C content in the leaves of transgenic plants could be due to refixation of photorespiratory CO_2 by PEPC and then Rubisco again can not be ruled out. The interesting question here is how transgenic plants manage to maintain a higher stomatal conductance. The mechanism underlying this phenomenon is not quite clear. However, it is conceivable that an increased expression of PEPC in the guard cells would allow more fixation of atmospheric CO_2 into organic acids such as malate, which is stored in the vacuole. Consequently, inorganic solutes such as potassium move from subsidiary or epidermal cells into guard cells for balance of charge. The accumulation of ions in the vacuole lowers the water potential of the guard cells, thereby stimulating the osmotic uptake of water and increasing turgor for opening of stomates.

Under photoinhibitory and photooxidative conditions, PEPC transgenic rice plants grown in the field also exhibit a superior photosynthetic performance than untransformed plants (Ku et al 2000). Under these stress conditions, PEPC transgenic plants are capable of maintaining a higher photosynthetic rate then untransformed plants. The intrinsic quantum yield of PSII, as measured by Fv/Fm, is less inhibited by full-sunlight treatment alone or by a combination of methyl viologen and full-sunlight treatment in PEPC transgenic plants, relative to the wild-type plants. Methyl viologen accepts electrons from PSI and generates oxy-radicals. PEPC transgenic plants are capable of dissipating excess light energy through photochemical and non-photochemical means, as demonstrated by the measurements of qP and qN, respectively, more effectively than untransformed plants. Taken together, these results indicate that PEPC transgenic plants are less susceptible to photoinhibition or photooxidation, which may contribute to the increased photosynthetic capacity. The basis for PEPC transgenic plants' superior ability to dissipate excess light energy is not known at present.

High-level expression of the maize PEPC in the leaves of transgenic plants (up to 18%) must reduce the relative amounts of other photosynthetic proteins, and yet these plants are capable of maintaining comparable or even a higher photosynthetic capacity than untransformed plants. It is quite possible that overexpression of the maize C4 PEPC in the transgenic rice plants may influence the activity or kinetics of other photosynthetic enzymes. This needs to be evaluated in relation to the photosynthetic performance of these transgenic plants. On a chlorophyll basis, Rubisco activities in untransformed and transgenic plants are similar (Ku et al 2000). Also, the $K_m(CO_2)$ of Rubisco is the same between the two plants (about $12 \, \mu M$ at $30 \, ^{\circ}C$). However, the V_{max} of Rubisco is two times higher in PEPC transgenic plants than in untransformed plants. In addition, carbonic anhydrase (CA) is almost three times higher in the transgenic plants. CA catalyzes the hydration of CO_2 to HCO_3^-. These results suggest that enhanced expression of the maize PEPC in rice may have altered the expression or activation state of other photosynthetic enzymes in the leaves. This metabolic adaptation warrants further investigation. In any case, increased CA activity may enhance fixation of atmospheric CO_2 via PEPC as HCO_3^- is the active CO_2 species for PEPC (Hatch 1987). The higher V_{max} of Rubisco in the leaves of transgenic plants may compensate for its lower amount (as a percentage) due to overexpression of the maize PEPC (Ku et al 1999) and thus allow the plants to maintain a high photosynthetic capacity.

PPDK transgenic rice plants

The photosynthetic performance of PPDK transgenic plants was evaluated using the segregation populations from four primary transgenic lines that exhibit high levels of PPDK (Ku et al 2000). Most of the PPDK transgenic plants exhibit a higher photosynthetic rate (up to 35%) than the wild-type plants, and the higher photosynthetic rates are associated with increased stomatal conductance and higher intercellular CO_2 concentration. Thus, as with PEPC transgenic plants, PPDK transgenic rice plants may also be able to maintain a higher internal CO_2 level due to increased stomatal conductance. Increased expression of PPDK in the guard cells may function to supply PEP, the substrate for PEPC, for synthesis of organic acids and stimulate stomatal opening (Schnabl 1981).

How the elevated PPDK may affect carbon and nitrogen metabolism in leaves of transgenic rice plants awaits further investigation. The effects of elevated expression of maize PPDK on carbon metabolism in transgenic potatoes (C3) has been reported recently (Ishimaru et al 1998). PPDK activities in leaves of transgenic potatoes are up to fivefold higher than those of untransformed control plants. Analysis of metabolites shows that PPDK activity in leaves is negatively correlated with pyruvate content and positively correlated with malate content. It is suggested that elevated PPDK activity in the leaf may lead to a partial function of C4-type carbon metabolism. However, the altered carbon metabolism does not have any significant effect on other photosynthetic characteristics in the transgenic potatoes.

Transgenic rice plants simultaneously expressing maize PEPC and PPDK

Since PEPC catalyzes the initial fixation of atmospheric CO_2 in the C4 pathway and PPDK catalyzes the conversion of pyruvate to PEP, overexpression of both enzymes simultaneously may enhance fixation of atmospheric CO_2 via PEPC. Using conventional cross-hybridization, we have integrated both PEPC and PPDK genes into the same plants from two independent homozygous transgenic plants. As detected by specific antibodies raised against the two maize enzymes, the amounts of the two enzymes in the F1 hybrids are about half of those in the parents (Ku et al 2000). The photosynthetic performance of the transgenic plants expressing varying amounts of the two maize enzymes was first evaluated in the segregation population from one of the F1 hybrids. As expected, the segregation population exhibits different combinations for the amounts of the two enzymes, with some having only the same basal amount as the wild-type plants (without the maize gene inserted) to twice the amount of the parental transgenic plants (homozygous with respect to the inserted maize gene). The activities of each enzyme are well correlated with the amounts of the protein among these plants. Hybrid transgenic rice plants expressing high levels of both PEPC and PPDK tend to have a high photosynthetic rate than untransformed plants, again due to a higher stomatal conductance and a higher intercellular CO_2. It is quite possible that overexpression of both enzymes further enhances the capability of the plants to synthesize organic acids in the guard cells and consequently the conductance of CO_2 into the leaf.

Future directions

In summary, our physiological results demonstrate that introduction of maize PEPC and PPDK into rice has the potential to enhance its photosynthetic capacity. PEPC transgenic rice plants are also more tolerant to photoinhibition and photooxidation under field conditions. This trait is important for rice productivity as early senescence of leaves, due to photoinhibition and photo-oxidation, often occurs in the field, which reduces grain filling. The performance of these transgenic plants under other stress conditions, such as drought, heat, chilling and mineral deficiency, needs to be evaluated in the future. The higher stomatal conductance exhibited by the transgenic rice plants implies that more water may be needed. However, this may not be a serious problem for paddy rice. On the other hand, an increased stomatal conductance may help cool the leaf canopy at high temperatures, as shown in wheat (Fischer et al 1998). A preliminary, small-scale field trial shows that the grain yield is about 10–20% higher in selected PEPC and 30–35% higher in selected PPDK transgenic rice lines and the hybrids between these two transgenic lines, relative to untransformed plants, in spite of a lower fertility (5%) in some of the transgenic plants. The increased yields are mainly due to increased tiller number. More field tests on a large scale are underway to confirm this. Also, whether this trait will be stably inherited in the following generations needs to be evaluated.

As one would expect, transgenic rice plants overexpressing maize PEPC, PPDK or both may not be capable of fixing large amounts of atmospheric CO_2 directly via the C4 pathway due to limited supply of substrates or further metabolism of reaction products. However, with the introduction of another key enzyme of the C4 pathway, NADP-ME in the chloroplast, a limited CO_2-concentrating mechanism, as exhibited by *H. verticillata*, may be achieved. Recently, we have obtained transgenic rice plants simultaneously expressing PEPC, PPDK and MADP-ME using cross-hybridization. These plants grow more rapidly and put out more tillers than the parental transgenic lines. Whether they are capable of concentrating CO_2 in the leaf and suppressing photorespiration is under investigation. Enhanced expression of other biochemical components of the C4 pathway, such as CA, NADP-malate dehydrogenase and adenylate kinase may allow the cycle to function more effectively. In this regard, the increased activities of CA and Rubisco in PEPC transgenic rice plants is worth noting; some related enzymes in the pathway may be induced or enhanced and it may not be necessary to genetically alter them simultaneously. For most efficient operation of the C4 pathway to concentrate CO_2 around Rubisco in the leaf the concomitant installation of Kranz leaf anatomy will be essential. Genes related to Kranz anatomy formation will have to be isolated and characterized before this can take place.

References

Agarie S, Tsuchida H, Ku MSB, Nomura M, Matsuoka M, Miyao-Tokutomi M 1998 High level expression of C_4 enzymes in transgenic rice plants. In: Garab G (ed) Photosynthesis: mechanisms and effects, vol V. Kluwer Academic, The Netherlands, p 3423–3426

Aoyagi K, Bassham JA 1984a Pyruvate orthophosphate dikinase of C_3 seeds and leaves as compared to the enzymes from maize. Plant Physiol 76:387–392

Aoyagi K, Bassham JA 1984b Pyruvate orthophosphate dikinase mRNA organ specificity in wheat and maize. Plant Physiol 76:278–280

Aoyagi K, Chua NH 1988 Cell-specific expression of pyruvate, Pi dikinase. Plant Physiol 86:364–368.

Blanke MM, Lenz F 1989 Fruit photosynthesis: a review. Plant Cell Environ 12:31–46

Bowes G, Salvucci ME 1989 Plasticity in the photosynthetic carbon metabolism of submersed aquatic macrophytes. Aquat Bot 34:233–249

Brown RH, Bouton JH 1993 Physiology and genetics of interspecific hybrids between photosynthetic types. Annu Rev Plant Physiol Plant Mol Biol 44:435–456

Douglas CJ 1996 Phenylpropanoid metabolism and lignin biosynthesis: from weeds to trees. Trends Plant Sci 6:171–178

Drincovich MF, Casti P, Andreo CS, Donahue R, Edwards GE 1998 UV-B induction of NADP-malic enzyme in etiolated and green maize seedlings. Plant Cell Environ 21:63–70

Ehleringer JR, Sage RF, Flanagan LB, Pearcy RW 1991 Climate change and the evolution of C_4 photosynthesis. Trends Ecol Evol 6:95–99

Fischer RA, Rees D, Sayre KD, Lu Z-M, Condon AG, Saavedra AL 1998 Wheat yield progress associated with higher stomatal conductance and photosynthesis rate, and cooler canopies. Crop Sci 38:1467–1475

Fushimi T, Umeda M, Shimazaki T, Kato A, Toriyama K, Uchimiya H 1994 Nucleotide sequence of a rice cDNA similar to a maize NADP-dependent malic enzyme. Plant Mol Biol 24:965–967

Gallardo F, Miginiac-Maslow M, Sangwan RS et al 1995 Monocotyledonous C_4 NADP-malate dehydrogenase is efficiently synthesized, targeted to chloroplasts and processed to an active form in transgenic plants of the C_3 cotyledonous tobacco. Planta 197:324–332

Gehlen J, Panstruga R, Smets H et al 1996 Effects of altered phospho*enol*pyruvate carboxylase activities on transgenic C_3 plant *Solanum tuberosum*. Plant Mol Biol 32:831–848

Gifford RM, Thorne JH, Hitz WD, Giaquinta RT 1984 Crop productivity and photoassimilated partitioning. Science 225:801–808

Hatch MD 1987 C_4 photosynthesis: a unique blend of modified biochemistry, anatomy and ultrastructure. Biochim Biophys Acta 895:81–106

Herrmann KM 1995 The shikimate pathway: early steps in the biosynthesis of aromatic compounds. Plant Cell 7:907–919

Hudspeth RL, Grula JW, Dai Z, Edwards GE, Ku MSB 1991 Expression of maize phospho*enol*pyruvate carboxylase in transgenic tobacco. Plant Physiol 98:458–464

Imaizumi N, Ku MSB, Ishihara K, Samejima M, Kaneko S, Matsuoka M 1997 Characterization of the gene for pyruvate,orthophosphate dikinase from rice, a C_3 plant, and a comparison of structure and expression between C_3 and C_4 genes for this protein. Plant Mol Biol 34:701–716

Ishimaru K, Ohkawa Y, Ishige T, Tobias DJ, Ohsugi R 1998 Elevated pyruvate, orthophosphate dikinase (PPDK) activity alters carbon metabolism in C_3 transgenic potatoes with a C_4 maize PPDK gene. Physiol Plant 103:340–346

Kogami H, Shono M, Koike T et al 1994 Molecular and physiological evaluation of transgenic tobacco plants expressing a maize phospho*enol*pyruvate carboxylase gene under the control of the cauliflower mosaic virus 35S promoter. Transgenic Res 3:287–296

Ku MSB, Kano-Murakami Y, Matsuoka M 1996 Evolution and expression of C_4 photosynthesis genes. Plant Physiol 111:949–957

Ku MSB, Agarie S, Nomura M et al 1999 High-level expression of maize phospho*enol*pyruvate carboxylase in transgenic rice plants. Nat Biotechnol 17:76–80

Ku MSB, Cho D, Ranade U et al 2000 Photosynthetic performance of transgenic rice plants overexpressing maize C_4 photosynthesis enzymes. In: Sheehy JE (ed) Redesigning rice photosynthesis to increase yield. Elsevier Science Publishers, Amsterdam, p 193–204

Magnin NC, Cooley CA, Reiskind JB, Bowes G 1997 Regulation and localization of key enzymes during the induction of Kranz-less, C_4-type photosynthesis in *Hydrilla verticillata*. Plant Physiol 115:1681–1689

Matsuoka M, Fukayama H, Tsuchida H et al 2000 How to express some C4 photosynthesis genes at high levels in rice. In: Sheehy JE (ed) Redesigning rice photosynthesis to increase yield. Elsevier Science Publishers, Amsterdam, p 167–175

Pinto ME, Casti P, Hsu TP, Ku MSB, Edwards GE 1999 Effects of UV-B radiation on growth, photosynthesis, UB-V-absorbing compounds and NADP-malic enzyme in bean (*Phaseolus vulgaris* L.) grown under different nitrogen conditions. J Photochem Photobiol B Biol 48:200–209

Reiskind JB, Madsen TV, van Ginkel LC, Bowes G 1997 Evidence that inducible C_4-type photosynthesis is a chloroplastic CO_2-concentrating mechanism in *Hydrilla*, a submersed monocot. Plant Cell Environ 20:211–220

Schaaf J, Walter MH, Hess D 1995 Primary metabolism in plant defense. Plant Physiol 108:949–960

Schnabl H 1981 The compartmentation of carboxylating and decarboxylating enzymes in guard cell protoplasts. Planta 152:307–313

Walter MH, Grima-Pettenati J, Feuillet C 1994 Characterization of a bean (*Phaseolus vulgaris* L.) malic enzyme. Eur J Biochem 224:999–1009

Winter K, Holtum JAM, Edwards GE, O'Leary MH 1982 Effect of low relative humidity on $\delta^{13}C$ value in two C_3 grasses and in *Panicum milioides*, a C_3–C_4 intermediate species. J Exp Bot 132:88–91

DISCUSSION

G.-L. Wang: In your transgenic plants is the grain quality decreased?

Ku: The size of the grain in our transgenic rice is comparable or slightly larger than that of the wild-type, but we haven't had a chance to examine the grain composition.

Beyer: You have shown a few very surprising effects that have to do with the induction of internal genes that even have an impact on plant architecture. However, the transformation itself is biochemically quite simple. Do you have any idea what the signalling molecule could be that provokes these interesting side effects?

Ku: When we put these maize genes in rice with high levels of expression, it may alter the metabolism. Gene expression and activity for many enzymes are known to be subject to metabolic regulation. In the case of photosynthetic enzymes, sugar molecules are involved.

Horton: Could you clarify why you think the photosynthetic rate increases? You show a rather small increase in the internal CO_2 in leaf as a result of the increased stomatal conductance. If you measure the A/Ci responses, is this increase in CO_2 level adequate to explain the increase in photosynthetic rate?

Ku: The increase in internal CO_2 is not small; in some transgenic lines there is about a 30 ppm increase. In C3 plants photosynthesis is highly limited by the ambient CO_2 level. A 10% increase in CO_2 is very significant.

Horton: You also showed some increase in Rubisco activity.

Ku: If you look at the protein profile, because of over-expression of PEP carboxylase or dikinase, some other photosynthetic enzymes in the leaf actually decrease a bit. The increase in Rubisco activation state may be a compensatory effect to maintain high rates of photosynthesis in these plants.

Nevill: What steps will you now take to try to understand the side effects you are causing in the transgenic plants? I ask from the point of view of being a potential 'ecoterrorist': it actually worries me that when you put a gene into a plant expecting it to do one thing, it actually does about 10 things that are completely different.

Ku: We need to go through all the related enzymes in the pathway, to look at their alterations. So far we know that carbonic anhydrase is being induced by the overexpression of PEP carboxylase and that Rubisco has a higher activation state. There may be some metabolic regulation of gene expression. This is very common for all the photosynthetic enzymes. As to how Rubisco acquires a higher activation state, I have no clues at present. This could be due to post-translational metabolic regulation. There are some earlier studies showing that some metabolites are involved in the activation of Rubisco.

Beyer: My belief is that since we have similar problems and results, metabolic profiling studies urgently need to be done. Has anything been set up for rice in this respect that regards the grain as a sack filled with chemicals?

Ku: Not to my knowledge. I think this is a fairly new approach.

Beyer: It is a new name for an old thing: you can call this a thorough analysis of all the compounds in a rice grain, which you should know about when you change them.

Khush: We have not looked at the chemical composition of the grain, but we should. From the observations of the general morphology we have detected many changes. Sterility is one, which occurs in many transgenic plants. In a few cases we have seen shortening of the panicle and reduction in number of grains. We have also seen changes in leaf architecture, but we have not done chemical analysis of the grain.

Okita: One of the things that we have learned from using knockouts and antisense approaches for metabolic engineering is that the plant is pretty plastic in terms of adjusting to these changes. The classic example is the knockout of the phosphate translocator, which is the main metabolite transporter that exchanges carbon between the chloroplast and the cytoplasm. You would think that if you knocked this out it would have drastic consequences on plant metabolism. But the plant survives nicely, because it utilizes a second transporter, a glucose transporter. It would be nice to use metabolic profiling techniques, and also RNA profiling techniques to see the changes in gene expression caused by the introduction of transgenes.

Horton: It seems rather surprising that a rice plant has found itself in a situation where its growth rate and yield has been held back by the inadequacy of what is a fundamental process in the plant — stomatal opening. It seems odd that the plant hasn't learned how to open its stomata a bit more and thereby grow faster. Have you any comments on this?

Ku: I have no answer. It is surprising why there is a fundamental difference between plants in terms of the synthesis of organic metabolites in the guard cells for stomatal regulation. Some plants appear to depend more on inorganic solutes while others depend more on organic solutes for building up turgor pressure in the guard cells, which promotes opening of stomates. From our results it seems that the enzymes relating to organic solute production in the guard cells are deficient in some plants such as rice. When you increase these enzymes in the guard cells, it results in an increased CO_2 diffusion into the leaf.

Leung: The plants have probably made a trade-off between how wide they open their stomata and some competing variable. One possibility is pathogens or something else enters through openings in the cuticle. Maurice Ku pointed out that it will be very important to check these transgenes under stress conditions.

Khush: Were the yield data you reported on an individual-plant basis or from a field-plot trial?

Ku: The first field trial consisted of plants growing in pots, but they were maintained in the net house. The second field trial is underway, and is taking place in the paddy, with a small patch of 50 plants per line. The differences observed in the field trials are consistent with those seen in the growth chambers under optimal conditions.

Khush: Is it a replicated trial?

Ku: No, just 50 plants per line and a few sublines for each transgenic line.

Okita: The increase in yield caused by the transgene is basically through the elevation of the number of tillers. As I understand it the new plant type selection criterion was to reduce the number of tillers per plant. Is this correct?

Khush: That is true. We are trying to reduce the tiller number and increase the number of grains per panicle. Rice variety IR72 has 100–120 grains per panicle, whereas the new plant type has about 200–250. The reason for this is that we needed to introduce more lodging resistance, and with too many tillers there is less lodging resistance.

Bennett: Have you had a chance to compare the tillering response of the wild-type and the transformant to spacing? The tiller number of the wild-type changes dramatically on crowding. Is the transgenic plant fixed at a higher tillering capacity, or does it still have a capacity to adjust on crowding? Are you revealing a mechanism by which crowding controls tiller number?

Ku: We haven't had a chance to address this. However, the tiller number is very consistent between the pot-grown and field-grown transgenic plants.

G.-L. Wang: In your Westerns you showed that different transgenic plants have different levels of enzyme activity. You say this is because of trangene copy number. How do you know this?

Ku: With primary transgenic plants, we assayed enzyme activity along with Western, Northern and Southern blot analyses to determine the effect of gene copy. For the hybrid study, we assayed the enzyme activity in the hybrid plants and their segregation populations, and then looked at the protein level by Western blot analysis and came up with a quantitative ratio of protein amount. In the homozygotes, the enzyme activity and protein amount are twice of those in the hybrid plants. This fits the classical segregation pattern.

G.-L. Wang: Did you see any correlation of the yield of these individual plants with the increase of enzyme activity?

Ku: These are segregation plants grown in a growth chamber in small pots. Now we are doing more segregation studies in the field, and we can test the relationship between enzyme activity and yield.

Leach: In the NADP-ME plants, which remain green past grain set, were more grains filled? I thought that one of the ideas that Gurdev Khush put forward was that there would be more grain fill if the plants remained green longer.

Ku: Those particular lines that overexpress NADP-ME look better in terms of fertility. It is rare to see empty seeds from these plants. One of the possible reasons for these plants being greener is that malic enzyme is very important for supplying NADPH, which is required for chlorophyll and secondary metabolite synthesis. We have carried out UV radiation studies with bean (Pinto et al 1999), and one of the major enzymes up-regulated is NADP-ME. The enzyme activities correlate with the amounts of UVB screening material. This enzyme has a side-effect not only on photosynthesis, but also in protecting the plants from stress conditions.

Elliott: Maurice Ku, in your paper you made it clear that *Hydrilla* was the model system upon which your transformations were based. Earlier, in an exchange with John Bennett, you flirted with the question as to whether a move towards full C4 plant status was realistic. I think you came to the conclusion that this was not a realistic target for the foreseeable future. Is this correct? If you think it might be, what would the strategy be?

Ku: It all comes down to money. If we have enough resources, I am optimistic that we could isolate the bundle sheath cell related genes. This is why we are interested in using Virginia Wolbot's maize transposon tagged mutants for screening such genes. Hopefully we will be able to find some C4 Kranz anatomy-deficient mutants. The other approach could take advantage of a very unique plant system: I am hopeful that someone can isolate these Kranz-related genes using an aquatic plant that occurs in Florida, called *Eleocharis vivipara*. This is a very interesting plant because under submerged conditions it is a C3 plant, but when it emerges both the biochemistry and Kranz anatomy of C4 plants are induced in the newly developed leaves within a few days. If we can do some kind of differential screening during this transitional period, we might find a gene that acts as a master switch for this process.

Matsumoto: Apparently, rice was the first cereal to branch off from the evolutionary tree, and that maize was the next. We know that rice has some of the genes involved in C4 photosynthesis, but these are not active. What do you think the role of these ancient genes were in rice?

Ku: Basically these genes encode protein/enzymes with similar functions as in C4 plants, except that they do not contribute significantly to overall photosynthetic CO_2 assimilation. Makoto Matsuoka has done quite a bit of work comparing the gene structure between rice and maize. There are some genes that have been highly conserved during evolution, such as pyruvate, orthophosphate and dikinase. The structure of the gene is almost identical between C3 rice and C4 maize. The only difference is in expression level: the promoter region for the gene in maize has been modified and the modification is not that big. It will be interesting to see whether we can turn on the endogenous genes in rice and convert it to C4 rice. I am pretty sure that there are some homologues in rice, even for the Kranz anatomy genes. In fact, if you look at the leaf anatomy of rice, there is reasonable bundle sheath cell differentiation. The problem is, there are not too many organelles, such as mitochondria and chloroplasts, associated with these cells.

Dong: In terms of evolution, where did C4 plants appear?

Ku: All C4 plants evolved from ancestral C3 plants fairly recently in warm and drought conditions. Jim Ehleringer at the University of Utah has good data showing that C4 plants evolved within the last 10 million years: they have multiple origins, and this transition occurred randomly in both dicots and monocots.

Dong: This would suggest that you can't really use the C3 plant enzymes.

Ku: In order to function in the C4 pathway, these enzymes usually modify their kinetic properties. For example, there is a co-evolution in terms of increased V_{max} and K_m for some C4 enzymes. This is because in C4 plants there is plenty of substrate for the enzyme. The enzyme doesn't have to work so hard in terms of affinity towards the substrate, as in C3 plants.

Khush: Wasn't there some evidence that some of the wild rice species, such as *Oryza rufipogon*, have an intermediate pathway between C3 and C4?

Ku: I am not aware of this and would like to look into this in the future. We have been working on C3/C4 intermediate species for quite some time now. There are two types. In *Panicum*, the intermediates have very good Kranz anatomy, but they don't have many C4 enzymes at all. They confine photorespiration in bundle sheath cells. Within these bundle sheath cells the photorespiratory CO_2 would be re-fixed by Rubisco. This is a modification or compartmentation of photorespiration; it has nothing to do with C4 biochemistry. On the other hand, if you look at the *Flavaria* C3/C4 intermediates, there is a co-evolution of Kranz anatomy formation and acquisition of a C4 biochemistry. Although these plants have well differentiated cell types, in terms of compartmentalization of the enzyme or the biochemical step, they do not strictly follow typical C4 behaviour. So their efficiency is not as high as in a typical C4 plant. It will be interesting to examine the nature of the intermediate state in this wild rice species.

Horton: In December 1999 there was a workshop at IRRI on prospects for developing C4 rice. It might be worth outlining some of the points that were made in terms of suggesting that it might not be as easy to engineer C4 photosynthesis into rice as Maurice is suggesting it could be. For C4 to work in terms of it giving the plant an advantage, it is absolutely essential that the enzymatic processes are highly regulated. Also, a critical factor seems to be the leak rate of the CO_2 that the plant is trying to concentrate. If the anatomy and the enzymology are not quite right, such that the CO_2 is leaking, the plant immediately loses the advantage because it has increased its quantum requirement. There was a great deal of concern about whether or not it would be possible to engineer everything such that all of this is exactly right. In the case of *Hydrilla*, a CO_2-concentrating mechanism is induced under quite strong CO_2 limitation. In some ways the situation is rather different from what you have in a rice plant, which is already photosynthesizing and growing quite quickly. Again, it is a question of the quantum requirement benefits that would result. The advantage of C4 is also absolutely dependent on the atmospheric level of CO_2. To engineer C4 in rice is likely to be a long-term project: by the time we get there it could well be that atmospheric CO_2 levels are high enough to make this redundant. The other point that was made is that when rice is grown under elevated CO_2, the yield benefit is not always evident. If there is no yield benefit from elevated CO_2, there certainly will be no benefit from C4.

Ku: I agree with your assessments on the difficulty of engineering the entire C4 syndrome in rice. However we can see benefits by expressing one or two C4 enzymes in rice, although it is not strictly related to the CO_2-concentrating mechanism of C4 photosynthesis. Another point to make is that even C4 plants can benefit from high CO_2. We recently published a paper on maize grown under high CO_2 conditions, which had a 20% increase in biomass (Maroco et al 1999).

References

Maroco JP, Edwards GE, Ku MSB 1999 Photosynthetic acclimation of maize to growth under elevated levels of carbon dioxide. Planta 210:115–125
Pinto ME, Casati P, Hsu TP, Ku MS, Edwards GE 1999 Effects of UV-B radiation on growth, photosynthesis, UV-B-absorbing compounds and NADP-malic enzyme in bean (*Phaseolus vulgaris* L.) grown under different nitrogen conditions. J Photochem Photobiol B 48:200–209

Increasing rice photosynthesis by manipulation of the acclimation and adaptation to light

Peter Horton, Erik H. Murchie, Alexander V. Ruban and Robin G. Walters

Department of Molecular Biology and Biotechnology, University of Sheffield, Western Bank, Sheffield S10 2TN, UK

Abstract. There are three important considerations in assessing the interaction of crop plants with light: (a) how does the plant respond to the light environment both in the short-term (regulation) and in the long-term (acclimation), (b) under what conditions are these responses inadequate, leading to photoinhibition, and (c) are the responses optimally adapted for maximum agricultural yield? Despite a wealth of knowledge about these processes in model plant species, it is impossible to predict how significant they are in influencing the yield of rice. Therefore, in collaboration with IRRI, we have undertaken a study of photoinhibition and photoacclimation of rice under field conditions. The results of this study are presented, along with an assessment of the implications for improvement of rice yield.

2001 Rice biotechnology: improving yield, stress tolerance and grain quality. Wiley, Chichester (Novartis Foundation Symposium 236) p 117–134

Light is the driving force for photosynthesis, plant growth and ultimately crop yield. Plants possess a large of number of features which maximize the harvesting of light, and its transduction into the biochemical products which fuel growth and development — and for crop plants, grain production. A large number of studies have correlated yield in terms of dry matter production with the amount of intercepted radiation, leading to formulation of indices such as radiation-use efficiency and radiation conversion factor (Mitchell et al 1999).

Yield is a function of biomass produced multiplied by harvest index. Since it is unlikely that any increases in harvest index above 0.6 are feasible, the increase in yield potential needed for rice will have to be derived from an increase in biomass. Similarly, since biomass produced is a function of leaf area index (LAI) and leaf photosynthesis, the yield increase will depend upon increased photosynthesis at the leaf level. In this article we will deal with the factors that seem to be important in determining the rate of photosynthesis in rice under field conditions.

117

Photosynthesis in a complex system in which the component reactions interact via regulatory and acclimation processes which tend to bring each one into balance in the face of challenges from the effect of environmental change. These processes are integrated into the developmental programming of the plant as it proceeds through vegetative growth, flowering, grain filling and senescence. A key question that has to be answered for a crop plant is whether these processes are optimized for agricultural yield. This question has to be answered separately for each crop, and indeed, for each particular environmental condition. For rice, each of the ecosystems (lowland rainfed, upland rainfed, deep-water and irrigated) and the wet and dry season have to be addressed separately. We may ask whether the adaptations of the plant are too weak, leading to stress under some conditions, or else too strong (inappropriate or conservative) leading to lost opportunities for higher photosynthesis and yield enhancement.

A principal feature of these processes is the interaction between the plant and light. The evolution of photosynthetic systems has not only led to the acquisition of processes which maximize conversion of sunlight into biochemical product, but also mechanisms which are a reaction to the fact that light is toxic, particularly in the presence of oxygen. The damaging effects of light derive from two kinds of processes — firstly, excited chlorophyll (Chl) may form triplet states, which can react with oxygen to form oxygen radicals. Secondly, the redox reactions induced by light can give rise to damage — in photosystem II oxidative damage to the pigments and proteins of the reaction centre occur thorough either from oxidized $P680^+$, or plastoquinol on the acceptor binding site (Barber 1995). The latter may proceed via radical oxygen species. This damage induces the proteolysis and disassembly of the reaction centre, which is replaced by new protein synthesis. Photosystem I can produce oxygen radicals on its acceptor side, which may also lead to damage to the proteins of the reaction centre (Terashima et al 1998). Production of oxygen radicals does not only induce deleterious effects on the reaction centres but may cause protein and lipid damage throughout the chloroplast. These damaging effects of light come under the term photo-inhibition, and as indicated above may include photo-oxidation.

When the Chl is associated with protein in the light-harvesting and reaction centres of the thylakoid membrane, photooxidative effects are minimized due to the presence of carotenoids which are dissipative to the Chl triplets. Special problems occur during the assembly and disassembly of the photosynthetic apparatus. In these circumstances the free or unconnected Chl or Chl precursors are very active oxygen radical producers if excited by light. For these reasons synthesis and breakdown of Chl has to be tightly coordinated with the synthesis and breakdown of the apoproteins.

Of course, much of the energy absorbed by Chl is used in photosynthesis, and this by itself 'protects' against some of the potentially deleterious effects of light. In

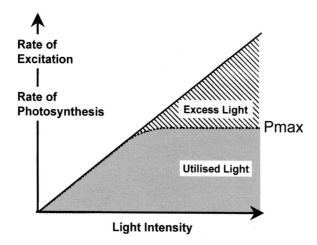

FIG. 1. Light saturation of photosynthesis and excess light. As irradiance increases photosynthesis reaches a saturation value, and light absorbed is then increasingly in excess.

this regard, the concept of *excess light* is a most useful one. The build-up of excitation energy and electrons in the reaction centre occurs when the rate of light absorption exceeds the rate of its utilization by electron transport. Since the dark reactions of photosynthesis have a finite capacity, the photosynthetic rate starts to saturate as the irradiance increases, until finally a ceiling rate (the P_{max}) is reached (Fig. 1). Therefore it is excess light that is potentially damaging. The extent of photodamage relates not to the absolute irradiance level but to the balance between irradiance and the capacity for photosynthesis (Horton 1985). A number of conditions may cause even very moderate levels of sunlight to be photoinhibitory (Fig. 2). As may be predicted, plants have evolved mechanisms to deal with excess light, and these fall into two categories:

- Constitutive: the presence of carotenoids to deal with triplet states; the presence of enzyme systems for getting rid of reactive oxygen species (although the level of these is modulated by factors that cause abiotic stress).
- Inducible: these are mechanisms that come into play only under excess light conditions (Fig. 2). There are two types: those induced rapidly, involving a change in function of existing molecules or structures; and those induced more slowly, involving a change in composition of the plant cell, frequently as a result of changes in gene expression. The former we shall group together under the heading 'photoprotection', and the latter we shall call 'photoacclimation'.

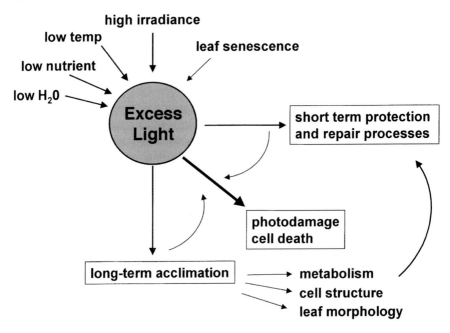

FIG. 2.   Causes of and responses to excess light. Any condition that upsets the balance between light input and metabolic demand gives rise to excess light. If unabated this will cause photodamage and cell death. Protection, repair and acclimation are plant responses to this challenge.

Photoprotection

Photoprotection may occur at several levels (Bjorkman & Demmig-Adams 1995). Firstly, in some plants (e.g. bean) the angle of the leaf relative to the direction of sunlight can be changed. When light is limiting, leaves may orientate themselves normal to the solar angle, maximizing absorption. Conversely, when light is in excess, the leaf angle may change so that the leaf surface is parallel to the direction of the sun, drastically reducing light absorption. Secondly, the absorptivity of a leaf can be modulated by the positioning of chloroplasts in the mesophyll cells — chloroplasts may be aligned such that self-shading occurs and total absorption is reduced.

Whereas the first two reduce the amount of light absorbed, excess light energy may be dissipated as heat. Of course all excess energy is dissipated as heat, but the important distinction here is the pathway for dissipation. Dissipation through triplet state formation and photooxidation is damaging. Photoprotective dissipation implies the creation or enhancement of a pathway for the rapid dissipation of Chl excited states such that in the steady state the accumulation of excitations is reduced, thereby decreasing photosystem II reduction and the frequency of triplet formation.

Reduction in the lifetime of Chl excited states gives rise to a decrease in the fluorescence yield — hence the induction of this dissipative pathway is detected by an increase in Chl fluorescence quenching. Because this quenching is not derived directly from a photochemical reaction, it is called non-photochemical quenching (NPQ). NPQ has been widely studied, is present in all plants, and is a biochemical regulatory mechanism that controls the level of excitation in the light harvesting system of photosystem II (Horton et al 1996).

When plants are exposed to light, the resulting NPQ is found to be heterogeneous. This heterogeneity was exposed by observation of the relation kinetics upon removal of light and also by the effects of various exogenous inhibitors (Horton 1996). Rapidly relaxing quenching is dependent upon the energization of the chloroplast by the thylakoid ΔpH and is called qE. More slowly relaxing NPQ is hard to ascribe in many situations — persistent ΔpH may cause long-lived qE, whereas other regulatory photoprotective processes can also give rise to sustained components of NPQ. One particularly confusing aspect is that damage to photosystem II also causes NPQ.

The qE component of NPQ has been the most widely studied. In addition to being dependent upon the ΔpH it is also controlled by the xanthophyll cycle. This cycle is the reversible de-epoxidation of a specific xanthophyll called violaxanthin into zeaxanthin. These pigments are associated with the light-harvesting complexes (LHCs) of the thylakoid membrane (Ruban et al 1999), and it is widely believed that it is within those associated with photosystem II (LHCII) that qE occurs (Horton et al 1996). A range of evidence indicates that qE arises from a protonation-induced conformation change in one or more of these proteins. The de-epoxidation state of the xanthophyll cycle carotenoids controls the pH-sensitivity of this process. The presence of carboxy-amino acids apparently essential for qE in the minor LHCII components CP29 and CP26 has led to suggestions that these are the principal sites of action of qE (Walters et al 1996, Pesaresi et al 1997). However, removal of these proteins by genetic manipulation does not prevent qE (J. Andersson, unpublished data), and attention has also focused upon the LHCII-related proteins such as ELIPs and PsbS as being possible sites of action of qE. The former proteins are induced under stress conditions and bind carotenoids, although it has yet to be shown that they specifically bind zeaxanthin. PsbS has also been reported to bind pigments, but again evidence that it is a site of zeaxanthin binding is absent. However, in PsbS-deficient mutants of *Arabidopsis*, rapidly relaxing qE is completely absent (Li et al 2000). Although the simplest explanation of this result is that PsbS is the unique site of qE it can not be discounted that it acts as a qE regulator, modulating the efficiency of other sites in LHCII.

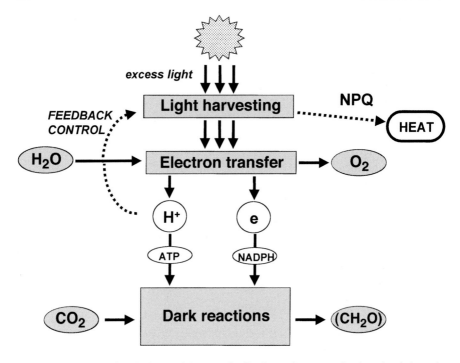

FIG. 3.   Non-photochemical quenching as a feedback regulatory mechanism that brings the rate of excitation of the photosynthetic reaction centres into balance with the demands of the metabolic dark reactions.

This form of NPQ has all the essential features required of a photoprotective regulatory mechanism. It is only induced when the linear part of the irradiance curve is passed, and its level increases strongly as saturation is approached. It generally reaches a maximum at about twice the irradiance needed to saturate photosynthetic carbon assimilation. It acts by feedback, the size of the ΔpH being the sensor of excess sunlight (Fig. 3). It is made very sensitive to rather small ΔpH changes by the fact that zeaxanthin is a positive effector of the proton-dependent quenching, with zeaxanthin synthesis itself drive by the acidification of the thylakoid lumen.

An important feature of NPQ is the variation in its extent when different species are compared (Johnson et al 1993). Plants adapted to low light tend to have a low NPQ capacity, whereas those which thrive under high or excess irradiance have largest capacity. Most cereal crops have rather high NPQ capacities as expected from their adaptation to open habitats. Some more exotic species with extreme tolerance to abiotic stress have unusually high NPQ capacities, which reach the maximum theoretical value of around 6 (Ruban et al 1993). In some species the potential for qE has been shown to vary according to growth conditions, with high capacity found in higher irradiances. There is no evidence to show which particular biochemical features of the LHCII system control this capacity. The involvement of PsbS in qE suggests that the level of this protein is the determinant of qE capacity. Other suggestions are that the other features of the thylakoid that are characteristic of sun-adapted or sun-acclimated plants are responsible — hence the decreased stacking, increased xanthophyll cycle pool size and lower LHCII concentration may allow greater structural flexibility and greater dynamic range of quenching (Horton 1999).

Genetic manipulation of photoprotection

The recent progress in understanding the molecular mechanism of NPQ presents new opportunities for genetic manipulation of plants to enhance, or indeed diminish this process. There is a possibility that the extent of NPQ may be controlled by manipulation of the expression level of the PsbS protein, although first more should be learned about not only how this protein participates in quenching, but also how its level is controlled. It remains to be shown for example whether this protein is expressed stoichiometrically with photosystem II or whether it forms a part of a stress-related group of genes. Genetic manipulation of the xanthophyll cycle is another attractive possibility for suppressing NPQ. There has been no exploration of genetic variation of NPQ in any crop plant, although variation has been found in the size of the xanthophyll cycle pool (Black et al 1995).

Photoacclimation

The composition of the chloroplast adjusts to the light environment (Anderson et al 1995). Under limiting light, high rates of photosynthesis are impossible, and the capacity of electron transport, ATP synthesis and carbon assimilation are reduced accordingly. Conversely, the content of Chl in the chloroplast increases, mostly due to the increased content of LHCII and LHCI. In contrast, under higher irradiance, high rates of photosynthesis are possible, and the plant invests in a higher content of electron transport components, ATP synthase and the enzymes of carbon assimilation, whereas the levels of LHC are reduced. These differences in composition give rise to the characteristic structural differences of the chloroplast in terms of the content of grana. These differences are found both when comparing plants adapted to sunny or shaded conditions, as well as, for many plant species, in those grown either under low or high irradiance. In addition to acclimation at the chloroplast level, leaf morphology may be very different under different light environments — in low light, leaves tend to be thinner because of a reduced number of chloroplast-containing mesophyll cells. This acclimation at the leaf level tends to reduce leaf chlorophyll content on a unit area basis, but may be offset by the higher chlorophyll content per chloroplast under these conditions. Across a range of plant species, differences in acclimation were described by the different strengths of leaf level and chloroplast level acclimation (Murchie & Horton 1997).

Acclimation is rationalized in terms of two related factors. Firstly, it optimizes the distribution of nitrogen into those components that are potentially limiting under different conditions. The two sides of photosynthesis, the light and dark reactions, are kept in balance. The second factor driving acclimation is the avoidance of photo-damage — by increasing photosynthetic capacity in high irradiance damage is of course avoided. It is remarkable that *Arabidopsis* grown over a wide range of irradiances (35–$600 \, \mu$moles m^{-2} s^{-1}) displayed a qP of around 0.9 and a photosystem II efficiency of over 0.75 at the growth irradiance — *acclimation prevents the occurrence of excess light*. However, a limit is reached in the achievable irradiance-dependent increase in P_{max} (see below), and the potential for photo-damage arises. Under these conditions further acclimatory changes occurs which are purely photoprotective by nature. These include an increase in the content of the xanthophyll cycle carotenoids, further reductions in chlorophyll content and an increase in stress-related proteins such as ELIPs. Associated with these changes is an increase in capacity for NPQ.

Many plants have the ability to dynamically respond to a change in irradiance during their growth cycle, in some cases by adjustments within existing leaves. Such changes may occurs within a few days of alteration in light conditions, and mean that plants can respond to factors such as alteration in climate and increased shading as a canopy develops.

Shading by other leaves includes not only a reduction in irradiance, but an enrichment in light not absorbed by Chl (i.e. increased far red). Alteration in spectral quality principally alters the content of photosystems II and I, and the interaction between the irradiance and spectral quality response can be quite variable between species (Murchie & Horton 1998). This feature of the shade environment also activates the phytochrome response, and induces stem elongation. In some plants this forms a shade avoidance response that is quite distinct from the type of acclimation discussed here.

Although a great deal is known about how chloroplast development is controlled at the level of gene expression, there is much less knowledge about how the various parts of the acclimation response are controlled. In unicellular algae there is extensive evidence that acclimation is controlled by the redox state of the acceptor side of photosystem II (Maxwell et al 1995, Escoubas et al 1995). The evidence for this occurring in higher plants is not so clear cut, but the similarity of the responses to low temperature and high irradiance has lent support to this kind of redox control. Expression of chloroplast-encoded genes has been shown to be under redox control (Pfannschmidt et al 1999). Analysis of a range of mutants of *Arabidopsis* has demonstrated a requirement for a blue light receptor for acclimation, and the involvement of various signal transduction elements in the process. Based on these data, a more complex model for the regulation of acclimation was proposed in which there is cross-talk with signalling from phytohormones and carbohydrates (Walters et al 1999).

The induction of photoprotective acclimation may well depend on different signals. Excess light could trigger a series of events: build up of carbohydrates, formation of reactive oxygen species, reduction of photosystem I acceptors. Each of these has been implicated in control of expression of photosynthetic genes, and could be involved in the high light response.

Genetic manipulation of photoacclimation

Because the signal transduction pathways involved in photoacclimation are so poorly understood, it is not yet possible to consider genetic manipulation to control the P_{max} of crop plants. However, using chlorophyll fluorescence analysis it is possible to isolate mutants which are altered in this process. Because at a given actinic irradiance, photosystem II efficiency differs for high-light and low-light acclimated plants, it is possible to screen mutants for the ability to change this parameter upon transfer between the two light environments. Using video imaging technology we have been able to isolate several mutants of *Arabidopsis thaliana* which can not acclimate to high light (R. G. Walters & P. Horton, unpublished data). Analysis of such mutants may well lead to the identification of target genes for manipulation in crop plants. Also, the same approach could be adapted to screen mutant populations of rice, and for genetic analysis of rice mapping populations.

Photoprotection and photoacclimation in the rice canopy

Photoprotection

Modern varieties of rice and other cereals have upright leaves and thus light may penetrate rather deeply even into mature canopies with high LAI. The upright posture of the upper leaves of rice gives rise to a characteristic diurnal pattern of photosynthesis in which one leaf surface received maximum irradiance around 9–10 am and the other surface being maximally irradiated in mid afternoon (Murchie et al 1999). At midday, photosynthetic activity of the upper leaves is rather low. At peak incident irradiance photosynthesis was light saturated. High levels of reduction of photosystem II and induction of NPQ were observed.

Most of NPQ was readily reversed indicating minimal photoinhibition. The erect leaves minimised the exposure to excess light. As the canopy matures a proportion of leaves become horizontal and in these cases NPQ was larger and more slowly reversible, the clearest evidence of excess light stress.

Photoacclimation

The light saturation of photosynthesis indicates that acclimation was incomplete. The saturation of the acclimation response was confirmed by imposition of shading in the field. Reduction of irradiance by over 60% did not reduce the P_{max}; instead there was a difference in Chl content and xanthophyll cycle pool size indicating a photoprotective response to full sunlight.

Lower leaves receive reduced irradiance, and were found to have a lower P_{max}. Part of this decline is consistent with that predicted from a photoacclimation response. However, the extent of the decrease was larger than expected and must also include a developmental change. In fact if P_{max} is recorded for rice leaves during their growth and development, a maximum P_{max} value is maintained only for a short period before it declines. Recycling of nitrogen is the reason for leaf senescence, and this is most likely the basis for the extensive decline in P_{max} of the lower leaves. Consistent with this view, there is a large decrease in level of Rubisco in lower leaves. The decreased P_{max} of lower leaves means that they too are also frequently light saturated and do not make full use of periods of high light arriving as sun flecks.

A particularly crucial phase of development of the rice crop is the period of reproductive growth. Here leaf senescence feeds the development of the panicle because uptake of new N is insufficient. A fine balance has to be struck since photosynthesis of the flag leaf is necessary to provide carbohydrate for grain filling, yet loss of N (from Rubisco and other leaf proteins) will tend to cause a decrease in photosynthesis. The increased tendency for photoinhibition, and the effects of crop management in which drainage of water may impose drought complicate this problem. Measurements of leaf composition during the grain-filling period reveal complex interactions symptomatic of the interplay between different regulatory mechanisms. P_{max}, which falls during panicle development in some cases, transiently increases in correlation with the rate of grain filling. Whilst loss of leaf protein proceeds continuously during this period, Rubisco is maintained, before sharply falling as grain filling is approaching completion.

The importance of leaf protein content as a source N for grain filling has been discussed (Horton 2000). It has been argued in fact that a high LAI is needed to enable enough N to be accumulated by the crop if the demands of grain yield are to be met (Sinclair & Sheehy 1999). LAI is frequently higher than that needed for canopy photosynthesis, and it has been argued that the benefits of erect leaf posture derive from the existence of such a high LAI — this allows the lower leaves to receive light for photosynthesis. In this context it is clear that the interplay between photoacclimation and leaf development are particularly important in terms of canopy architecture (Horton 2000, Horton & Murchie 2000). If the level of leaf N could be enhanced, LAI could be reduced and respiratory losses decreased. In this regard it is particularly interesting that the NPT (new plant type) rice has very large Rubisco content, in excess of that needed to support P_{max}.

Summary of factors affecting rice leaf development

Rice leaf development is influenced by several factors: light conditions; plant development (new leaves, grain); canopy factors (self-shading); and supply of nutrients (N). These responses are determined by genetic adaptation of rice during its evolution, and may give rise to values of P_{max} and N content which are not optimized for maximum agricultural yield.

Implications for wet and dry season rice, and marginal habitats

For the irrigated rice crop in the dry season, a major constraint on radiation conversion appears to be the light saturation of photosynthesis. Crop improvement would result from an increase in P_{max}, and this might be achieved if we understood what controls the photoacclimation process (see above). This is perhaps a more promising approach than trying to identify a single limiting step and increasing its level, since it is highly likely another limitation would be reached. A second approach is to improve canopy structure, to reduce the proportion of leaves that are light-saturated. A third and very important approach is to reduce losses from photorespiration, but this is outside the scope of this article, and has been covered in detail elsewhere (Sheehy & Peng 2000).

It is particularly important to consider the differences between the geographical location of the rice crop and also the climatic season. The wet season presents very different constraints from the dry season for irrigated rice. Temperatures are lower, and levels of radiation significantly reduced by cloud cover. Hence increasing the ceiling level of P_{max} is not important under these conditions — better performance in reduced irradiance might be more significant. Yield data for rice varieties in the wet and dry season revealed that the yield advantage of certain NPT rice over IR72 was much more significant in the wet season (Peng et al 1999). It is interesting therefore that we have shown that P_{max} of NPT is higher than IR72 when grown under shaded conditions whereas the opposite is the case in full sunlight (Horton & Murchie 2000). Hence for the wet season, varieties that do not *strongly* acclimate to reduced irradiance may be preferred — a reduced P_{max} would prevent exploitation of periods of high light arriving as sunflecks.

In terms of photoprotection it has been argued that this and other metabolic regulatory mechanisms perhaps reflect the conservative aspect of the evolution of rice, and its adaptation to low light (Horton 1994, 2000). Stability towards environment change and competition for resources are more important that growth rate or yield, and many regulatory processes could be removed without deleterious effect under good agricultural conditions. In fact, these processes may restrict maximum growth rate, and their removal may promote enhanced yield.

In harsher habitats, such as in rainfed lowland rice, abiotic stress arising from drought or submergence is a significant factor affecting yield (Wade et al 1999). There have been no studies of photoinhibition and photoprotection of rice under these conditions. However, since the most severe stress is during the reproductive phase, photoinhibition is expected to have an impact on grain filling. Breeding new varieties with enhanced photoprotection and acclimation to excess light may be beneficial for rice growing in these environments.

Conclusion

In summary, the reactions of all plants, including rice, to the light environment is one of the dominating forces in determining form and function. Manipulation of these reactions offers many opportunities for increasing crop yield. Data from field experiments indicates that photosynthesis in rice is not optimally adapted to tropical conditions and that excess light stress may lead to photoinhibition under some conditions. There are also restrictions on the expression of photosynthetic capacity both in the short term and developmentally. Therefore, there are opportunities for crop improvement from modification of leaf (and canopy) development. It is important that we first learn more about flag leaf photosynthesis in relationship to supply of C and N to the grain, and it is vital to consider different strategies for yield enhancement for different climates and ecosystems. Non-invasive assay of photosynthetic performance by chlorophyll fluorescence analysis offers a good methodology for identifying important genes by screening mutant collections and mapping populations.

References

Anderson JM, Chow WS, Park Y-I 1995 The grand design of photosynthesis: acclimation of the photosynthetic apparatus to environmental cues. Photosynth Res 46:129–139

Barber J 1995 Molecular basis of the vulnerability of photosystem II to damage by light. Aust J Plant Physiol 22:201–208

Black CC, Tu Z-P, Counce PA, Yao P-F, Angelov MN 1995 An integration of photosynthetic traits and mechanisms that can increase crop photosynthesis and grain production. Photosynth Res 46:169–175

Björkman O, Demmig-Adams B 1995 Regulation of photosynthetic light energy capture, conversion, and dissipation in leaves of higher plants. In: Schulze ED, Caldwell MM (eds) Ecophysiology of photosynthesis. Springer, Berlin, p 17–47

Escoubas JM, Lomas M, LaRoche J, Falkowski PG 1995 Light intensity regulation of *cab* gene transcription is signalled by the redox state of the plastoquinone pool. Proc Natl Acad Sci USA 92:10237–10241

Horton P 1985 Interactions between electron transport and carbon metabolism. In: Barber J, Baker NR (eds) Photosynthetic mechanisms and the environment. Elsevier Science Publishers, Amsterdam (Topics in Photosynthesis vol 6) p 135–187

Horton P 1994 Limitations to crop yield by photosynthesis. In: Cassman KG (ed) Breaking the yield barrier. IRRI, Los Baños, Philippines (Proceedings of a workshop on rice yield potential in favorable environments, Los Baños 1993) p 111–115

Horton P 1996 Nonphotochemical quenching of chlorophyll fluorescence. In: Jennings RC, Zucchelli G, Getti F, Colombetti G (eds) Light as an energy source and information carrier in plant physiology. Plenum Press, NY (Proceedings of a Nato Advanced Study Institute, Voltera, Italy 1994) p 99–111

Horton P 1999 Are grana necessary for the regulation of light harvesting? Aust J Plant Physiol 26:659–669

Horton P 2000 Prospects for crop improvement through the genetic manipulation of photosynthesis: morphological and biochemical aspects of light capture. J Exp Bot 51:475–485

Horton P, Murchie EH 2000 C4 photosynthesis in rice: some lessons from field studies of rice photosynthesis. In: Sheehy J, Peng S (eds) Redesigning rice photosynthesis to increase yield. Elsevier Science Publishers, Amsterdam, p 127–144

Horton P, Ruban AV, Walters RG 1996 Regulation of light harvesting in green plants. Annu Rev Plant Physiol Plant Mol Biol 47:655–684

Johnson GN, Young AJ, Scholes JD, Horton P 1993 The dissipation of excess excitation energy in British plant species. Plant Cell Environ 16:681–686

Li X-P, Björkman O, Shih C et al 2000 A pigment-binding protein essential for regulation of photosynthetic light harvesting. Nature 403:391–395

Maxwell DP, Laudenbach DE, Huner NPA 1995 Redox regulation of light harvesting complex II and cab messenger-RNA abundance in *Dunaliella salina*. Plant Physiol 109:787–795

Mitchell PL, Sheehy JE, Woodward FI 1999 Potential yields and the efficiency of radiation use in rice. IRRI, Los Baños, Philippines (IRRI Discussion Paper Series 32)

Murchie EH, Horton P 1997 Acclimation of photosynthesis to irradiance and spectral quality in British plant species: chlorophyll content, photosynthetic capacity and habitat preference. Plant Cell Environ 20:438–448

Murchie EH, Horton P 1998 Contrasting patterns of photosynthetic acclimation to the light environment are dependent on the differential expression of the responses to irradiance and spectral quality. Plant Cell Environ 21:139–148

Murchie EH, Chen Y-Z, Hubbart S, Peng S, Horton P 1999 Interactions between senescence and leaf orientation determine *in situ* patterns of photosynthesis and photoinhibition in field grown rice. Plant Physiol 119:553–564

Peng S, Khush G, Visperas R, Evangelista A 1999 Progress in increasing grain yield by breeding a new plant type. In: IRRI Program Report for 1998. IRRI, Los Baños, Philippines, p 7–9

Pesaresi P, Sandonà D, Giuffra E, Bassi R 1997 A single point mutation (E166Q) prevents dicychlohexylcarbodiimide binding to the photosystem II subunit CP29. FEBS Lett 402:151–156

Pfannschmidt T, Nilsson A, Allen JF 1999 Photosynthetic control of chloroplast gene expression. Nature 397:625–628

Ruban AV, Young A J, Horton P 1993 Induction of non-photochemical energy dissipation and absorbance changes in leaves; evidence for changes in the state of the light harvesting system of photosystem II *in vivo*. Plant Physiol 102:741–750

Ruban AV, Lee PJ, Wentworth M, Young A J, Horton P 1999 Determination of the stoichiometry and strength of binding of xanthophylls to the photosystem II light harvesting complexes. J Biol Chem 274:10458–10465

Sheehy J, Peng S (eds) 2000 Redesigning rice photosynthesis to increase yield. Elsevier Science Publishers, Amsterdam

Sinclair TR, Sheehy JE 1999 Erect leaves and photosynthesis in rice. Science 283:1456–1457

Terashima I, Noguchi K, Itoh-Nemoto T, Park YM, Kubo A, Tanaka K 1998 The cause of PSI photoinhibition at low temperatures in leaves of *Cucumis sativus*, a chilling-sensitive plant. Physiol Plant 103:295–303

Wade L J, Fukai S, Samson BK, Ali A, Mazid MA 1999 Rainfed lowland rice: physical environment and cultivar requirements. Field Crops Res 64:3–12

Walters RG, Ruban AV, Horton P 1996 Identification of proton-active residues in a higher plant light-harvesting complex. Proc Natl Acad Sci USA 93:14204–14209

Walters RG, Rogers J JM, Shephard F Horton P 1999 Acclimation of *A rabidopsis thaliana* to the light environment: the role of photoreceptors. Planta 209:517–527

DISCUSSION

Mazur: Are your photoacclimation mutants tagged genes? What is the gene?

Horton: The photoacclimation mutants we have isolated are tagged genes. The gene for one mutant we have characterized is in the *A rabidopsis* database and has no known function. It seems to be conserved in photosynthetic organisms in the sense that this gene is also present in *Synechocystis*. We have sequenced it and the predicted protein structure suggests that it is a membrane protein. Apart from this we know nothing else.

Mazur: Did you only find one type of mutant, or were there others?

Horton: We have many mutants. In some of them we pick up things we would not have anticipated. For example, one of the mutants we picked up was actually a mutant in the phosphate translocator. The explanation as to why we picked this up is that this mutation actually restricts photosynthetic rate under high light. We are trying to find proteins involved in the signalling process. This sort of screening technology is applicable to all sorts of other environmental influences. Because of the way in which the photosynthetic system is connected, almost anything that affects different parts of the process can be detected in terms of an effect on the chlorophyll fluorescence. Because you can screen so quickly with chlorophyll fluorescence, you can design relatively smart screens in this way.

Khush: The light intensity is higher under tropical conditions. Is that why we get higher yields in the temperate areas?

Horton: One would be tempted to say yes. That statement about rice not being optimally adapted relates to a number of things. It relates to temperature in relation to photorespiration and to irradiance. I am cautious about saying that the lighting conditions in a temperate environment are less severe than a tropical environment.

Bennett: The textbook written by Yoshida (1981) has a diagram that shows light intensities throughout the year at different sites. The light intensity during grain filling is higher at Los Baños than at any other site. It is during the vegetative stage that other parts of the world experience their highest light intensity. In the irrigated tropics, where the harvest is in April or May, the highest light intensity is experienced during grain filling, when 70% of the carbohydrate for the grain is produced.

Horton: That is interesting. This period of grain filling is the most important process to address in terms of its relationship with photosynthesis. As you say, 70% or more of the carbon is coming from photosynthesis in the flag leaf. Under these conditions, the environmental conditions are likely to be the most severe in terms of light. Also, during this period the fields are being drained to some extent, so the supply of water is perhaps not optimum as it was earlier on. There is also the problem that the plant might be losing photosynthetic capacity because it is losing protein from the leaf to fill the grain. It is a very finely balanced process. The genetic manipulations that I know you are engaged in to try to delay leaf senescence are of course very interesting in this regard, in terms of how beneficial delaying senescence will be. It is a question of getting the balance right, in preserving photosynthesis but also making sure that materials can be mobilized for the grain filling.

Ku: In temperate conditions, there may still be bright sunlight but the ambient temperature tends to be lower. Would you expect some degree of photoinhibition under such conditions?

Horton: It would depend how low the temperature is. There have been studies of photoinhibition in maize in the springtime in the UK (Long et al 1994). At this time, relatively bright sunlight and low morning temperatures are common. Photoinhibition was detected that had a lasting effect on the crop. In other words, the plants did not recover completely from this short period of photoinhibition during that stage of growth. This question of whether photoinhibition occurs depends on balance of the irradiance compared with other factors that are limiting. If there is very strong irradiance and slightly cool temperatures, this would be photoinhibitory as much as lower irradiance and very low temperatures.

Ku: You showed data that relative to IR72, the Rubisco content in the NPT rice is more than doubled. In terms of carbon gain by photosynthesis, however, there is not much improvement. How do you explain this in terms of nitrogen investment? You also alluded to the suggestion that it is important for the plant to have high nitrogen stored as Rubisco, which later degrades and supplies amino acids to the grain for synthesis of seed storage protein. It seems to me that these plants must be poor in nitrogen utilization efficiency.

Horton: It is clear that these plants have excess Rubisco. In other words, the rate of photosynthesis in NPT rice is not significantly higher than the IR72, despite the double Rubisco content. It could be argued that this is inefficient. But there is another way of looking at this. John Sheehy at IRRI has an interesting rationale about leaf nitrogen content (Sinclair & Sheehy 1999) — it has been suggested that the leaf area index of many crops is higher than it should be in terms of interception of light. One purpose of having a high LAI is to provide sufficient nitrogen in the aerial parts of the plant that can feed grain filling. If you accept this view, the NPT could be exactly what we are looking for: we want a plant with a high level of nitrogen, and therefore it wouldn't need as large a leaf area in order to feed the nitrogen requirement for the grain. With this, we could start to think about the way in which the canopy architecture could be redesigned. The only reason for having erect leaves is to make sure that light gets down to the lower leaves, such that these leaves are not totally parasitic. It is an interesting rationale. It was the observation we made about the high Rubisco content that led me into the conversation with John that I had about a year ago.

Khush: You say that there are opportunities for modification of leaf development. Which modifications do you think would be useful?

Horton: Leaf development needs to be optimized for radiation conversion (Horton 2000). We should try to ensure that the leaf doesn't reach the acclimation ceiling. If we can understand more about what determines the development of P_{max} in the leaf, there is no reason why we couldn't have a higher P_{max}. It is not a problem of manipulation of particular enzymes in the conventional way; it is a question of manipulating the developmental process that leads to having a particular level of enzyme. There is scope for having a higher P_{max} in a leaf. We can clearly have a higher nitrogen content in the leaf. We should also think about the response to low light. The particular variety of NPT that we looked at retained a high photosynthetic capacity in low light. This variety is in fact the best performing variety in the wet season, when irradiance is lower.

Khush: The yield potential of the NPT is much higher than IR72 in the wet season than it is in the dry season.

Beyer: You mentioned briefly the xanthophyll cycle. Is there any evidence, other than just correlation, that explains how this should help in adaptation to different light qualities? Is anything known on a molecular level?

Horton: We spent a lot of time trying to understand the molecular mechanism by which the xanthophyll cycle controls non-photochemical quenching (Ruban & Horton 1999). It has been a very controversial topic. We can show that with isolated light-harvesting proteins of different kinds *in vitro*, when zeaxanthin is bound to them there is non-photochemical quenching. Using various spectroscopic indicators we see evidence of structural change in the protein. We believe that when the zeaxanthin molecule binds to the protein, the protein changes its conformation, forming quenching states in the chlorophyll molecules that bring about the dissipation of energy. So the information that is available is not purely correlative at the leaf level, as it was in the literature until perhaps six years ago. Essentially this family of light-harvesting proteins are all similar, and they all behave in a rather similar way. One thing we don't know is which particular protein is the exact site, or indeed if there is more than one site. There is recent interest in one of the LHC-related proteins called PsbS. It could be that this is the protein that zeaxanthin binds to in the thylakoid membrane *in vivo* and acts as the quenching site.

Parker: What are the major mechanisms that respond to the redox imbalance that is created with excess light? How do they feed back into the photosynthetic machinery?

Horton: There is obviously an immense amount of interest in redox control. In terms of the short-term immediate responses to a change in light level, there is a very well characterized mechanism that John Bennett was involved in elucidating many years ago. In the reduced state of the electron transport chain, one of the light-harvesting proteins is phosphorylated, causing it to be detached from the photosynthetic reaction centre. This appears to be a process that responds to light quality, rather than light intensity. In photosynthetic acclimation, there is compelling evidence that at least part of the signal transduction process is a redox control. This appears to be also mediated through the redox states of the plastoquinone pool and/or the associated cytochrome complex. It has been shown that the expression of chloroplast-encoded genes is controlled by the redox states of the pool (Pfannschmidt at al 1999). In general, there is a hypothesis that the way in which acclimation works is to sense the redox state: that excess light results in excess electrons, and this would then trigger changes in gene expression in various compartments. This acclimation process occurs not just in the chloroplasts, but also at other locations in the leaf. There is also interaction between pathways, such as the central processing units in sugar signalling, and plant hormones. We looked at acclimation in different signal transduction mutants of *Arabidopsis*, and found that acclimation to light is disrupted in one of these signal transduction mutants (Walters et al 1999). So there is cross-talk between the redox control in the chloroplast and the controls involved in sugar signalling. Of course, carbohydrate is something that also builds up under conditions of excess light.

Parker: So would your mutant screens pick up defects in the redox signalling system?

Horton: We should pick up defects in the redox signalling system using this strategy, but it could be that there are other ways of being more selective for redox control. One thing we have been looking at is the induction of acclimation by DCMU, which has a specific effect on the redox state of the plastoquinone pool. In this way, we could screen for mutants with defects in the redox signalling alone.

References

Horton P 2000 Prospects for crop improvement through the genetic manipulation of photosynthesis: morphological and biochemical aspects of light capture. J Exp Bot 51:475–485

Long SP, Humphries S, Falkowski P 1994 Photoinhibition of photosynthesis in nature. Annu Rev Plant Physiol Plant Mol Biol 45:633–662

Pfannschmidt T, Nilson A, Allen JF 1999 Photosynthetic control of chloroplast gene expression. Nature 397:625–628

Ruban AV, Horton P 1999 The xanthophyll cycle modulates the kinetics of non-photochemical energy dissipation in isolated light harvesting complexes, intact chloroplasts and leaves. Plant Physiol 119:531–542

Sinclair TR, Sheehy JE 1999 Erect leaves and photosynthesis in rice. Science 283:1456–1457

Yoshida S 1981 Fundamentals of rice crop science. International Rice Research Institute, Manila, Philippines

Walters RG, Rogers JJM, Shephard F, Horton P 1999 Acclimation of *Arabidopsis thaliana* to the light environment: the role of photoreceptors. Planta 209:517–527

Increasing rice productivity and yield by manipulation of starch synthesis

Thomas W. Okita*, Jindong Sun*†, Chotipa Sakulringharoj*, Sang-Bong Choi*, Gerald E. Edwards†, Chikako Kato‡ Hiroyuki Ito‡ and Hirokazu Matsui‡

*Institute of Biological Chemistry, and †School of Biological Sciences, Washington State University, Pullman, WA 99164, USA and ‡Department of Applied Bioscience, Graduate School of Agriculture, Hokkaido University, Kita-9, Nishi-9, Kita-ku, Sapporo 060-8589, Japan

Abstract. Plant productivity and yield are dependent on source–sink relationships, i.e. the capacity of source leaves to fix CO_2 and the capacity of developing sink tissues and organs to assimilate and convert this fixed carbon into dry matter. Studies from our laboratories as well as others have demonstrated that rice productivity and yield are mainly sink-limited during its development because of limited capacity to utilize the initial photosynthetic product (triose phosphate). This limitation in triose phosphate utilization, evident at both the vegetative and reproductive stages of rice development, may be associated with limited capacity for carbohydrate synthesis in rice leaves (which are poor accumulators of starch) or feedback due to limited sink strength of developing seeds. Strategies in improving triose phosphate utilization by enhancing starch production in leaves and developing seeds by the expression of engineered genes for ADP glucose pyrophosphorylase, a key regulatory enzyme of starch biosynthesis, are discussed.

2001 Rice biotechnology: improving yield, stress tolerance and grain quality. Wiley, Chichester (Novartis Foundation Symposium 236) p 135–152

The world population is estimated to increase by about 2 billion in the next 50 years (Waterlow et al 1998). Such population growth will exceed present capacity for food production and will place a further strain on existing water resources and available arable land. In the 1960s and 1970s, the widespread usage of chemical inputs, fertilizers, pesticides and fungicides, as well as the implementation of improved agricultural practices have resulted in a significant jump in crop productivity and yields. In more recent years, annual increases in crop yields and production have only gradually increased suggesting that the genetic potential for maximum crop yields is close to being attained. To meet this challenge, new strategies to elevate crop productivity and yields must be devised to accommodate the growing demand for food. Such strategies will include genetic modifications to raise the ceiling for maximum yield (harvest index) and to increase yield potential under stressful environments.

Plant productivity and crop yields are dependent on source–sink relationships, i.e. the capacity to photosynthesize and fix CO_2 by source leaves and the capacity to assimilate this fixed carbon by sink tissues and organs. Source leaves capture light energy and fix CO_2 to produce sugars, starch, amino acids and other metabolites. Sugars and amino acids are exported from the source leaves and transported to developing sink tissues, e.g. young developing leaves and new root tissue, which utilize these basic precursors for growth and development. Because of the fundamental importance of primary CO_2 assimilation, the photochemical and biochemical activities of the source leaves play a dominant role in controlling plant productivity. Plant productivity in cereal crops is also influenced by the capacity of reproductive sink tissue to uptake and assimilate photosynthate produced by source leaves or re-converted from storage reserves (Ho 1988, Schaffer & Zamski 1996). The contributions of the source and sink tissues are not equivalent during the developmental cycle of the plant. During the vegetative phase of growth where numerous sink tissues such as developing leaves, axillary buds, roots etc. are being formed, overall productivity is more likely dictated by the extent of photosynthesis (Baker et al 1990, Board & Tan 1995, Hocking & Meyer 1991, Imai et al 1985). In contrast, during the reproductive stage of the plant, e.g. in cereal crops, the capacity of sink organs to utilize photoassimilates may control the extent of productivity (Chen & Sung 1994, Hocking & Meyer 1991, Rowland-Bamford et al 1990).

The extent of photosynthesis is governed by the availability of light and CO_2. Under low light, photosynthesis is limited mainly by the harvesting of solar energy and generation of assimilatory power. Under high light, moderate to high temperature, high stomatal resistance and/or high diffusive resistance, and high Rubisco, photosynthesis becomes limited by CO_2 availability (Leegood & Edwards 1996, Sharkey 1985, Sharkey et al 1995). Under CO_2- and light-saturated conditions, photosynthesis is controlled by processes that convert triose phosphate into carbohydrates especially sucrose (Sage 1990, 1994, Sims et al 1998, Stitt 1986, 1996). When limitations exist in utilizing triose phosphate produced during photosynthesis for synthesis of carbohydrates (in source or sink tissues), feedback/down-regulation on production of triose phosphate occurs. One general form of feedback in the short-term can occur through change in the levels of metabolites which can effect catalysis of key enzymes of the C3 cycle and carbohydrate synthesis, and is associated with build-up of organic phosphates, and photosynthesis being limited by inorganic phosphate (P_i) (Schaffer & Zamski 1996, Sharkey et al 1986, Stitt et al 1987). In the longer term down-regulation of photosynthetic gene products, e.g. Rubisco production, can occur (Jang et al 1997, Jang & Sheen 1997, Koch 1996, Lee & Daie 1997, Sheen 1994,

Smeekens & Rook 1997). Thus, the capacity to utilize triose phosphate for carbohydrate synthesis in leaves or sink tissue can establish an upper limit for the maximum rate of photosynthesis (Sage 1990, Sharkey 1990, Sharkey et al 1995).

Evidence that rice is sink-limited during the reproductive stage

Because of their importance as a food source, rice and the other short grain cereals such as wheat and barley have been studied extensively with regard to source–sink relationships. Moreover, the short grain cereals are an excellent experimental system as the bulk of the total carbon (60–90%) accumulated in developing seeds is produced from photosynthesis that occurs during heading and seed development, and much of this photosynthate is produced by the flag leaf (Yoshida 1981).

The role of photosynthesis in determining total biomass and the number of reproductive organs of short grain cereals is supported by CO_2 enrichment experiments. Under enriched CO_2 conditions, rice plants exhibit higher photosynthetic rates which lead to an increased number of tillers and reproductive structures (Baker et al 1990, Chen & Sung 1994, Imai et al 1985, Rowland-Bamford et al 1990, Ziska & Teramura 1992). These plants produce a larger yield of seeds than plants grown under normal atmospheric conditions. Except for one study (Ziska & Teramura 1992), seed weight per unit total biomass (harvest index) remain unchanged. Therefore, enhancement in photosynthesis and photosynthate under higher levels of CO_2 in rice, however, does not result in higher grain weight because of the inability of developing seeds to convert this photosynthate into dry matter (starch) (Chen & Sung 1994). In fact, growth under increasing levels of CO_2 above current levels resulted in a large linear decline in Rubisco content, indicating that there is strong acclimation and feedback effect on source activity due to inability of sinks to utilize additional photosynthate (Rowland-Bamford et al 1991). Hence, productivity at this stage of plant development is governed by the sink tissue (developing seeds).

Loss of sensitivity of photosynthesis to changes in CO_2 and O_2
suggests rice is more sink-limited compared to many other C3 plants

To further evaluate whether rice productivity was affected by sink strength we studied the photosynthetic response of rice plants at the reproductive stage. The effects of temperature, light, and changes in CO_2 and O_2 concentrations, on rates of CO_2 exchange and photosystem II activity were measured in leaves of rice (Taipei 309, japonica type) with plants grown at a photosynthetic photon flux density (PPFD) of 1000 μmol m^{-2} s^{-1} and 26 °C day/24 °C night (Choi et al 1998, Sun et al 1999a, Winder et al 1998). Even under moderate conditions of temperature

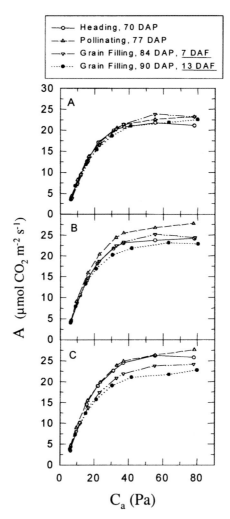

FIG. 1. CO_2 response curves for photosynthesis in the flag leaf of rice at heading and during grain filling. Photosynthesis as measured by CO_2 assimilation was measured with three separate plants (panels A, B and C) at a PFD of $700\,\mu$mol quanta m^{-2} s^{-1}, vapour pressure deficit of 3–4 mbars and a temperature of 26 °C. Reproduced with permission from Winder et al (1998).

(26 °C), light and vapour pressure deficit, photosynthesis of the flag leaf saturates at near ambient CO_2 levels especially under conditions that favour high stomatal conductance (Fig. 1). This saturation of CO_2 assimilation is not due to limitation of carboxylation capacity of Rubisco. The maximum rate of CO_2 assimilation observed ($22\,\mu$mol m^{-2} s^{-1}) was only about half of the theoretical $V_{c(max)}$ estimated for Rubisco ($42.3\,\mu$mol m^{-2} s^{-1}). This saturation at near ambient CO_2

levels is atypical for C3 species, which normally require two to three times higher ambient levels. Under elevated short-term CO_2 enhancement, saturation of photosynthesis in rice at near ambient CO_2 partial pressure indicates a limitation outside of the C3 cycle. This view is supported by the observation that rice shows a much lower stimulation of photosynthesis than predicted based on the enzyme kinetic properties of Rubisco when O_2 is reduced (Winder et al 1998). This lower O_2 sensitivity than predicted by modelling based on Rubisco properties (Sage & Sharkey 1987) suggests a limitation in triose phosphate utilization. Limitations on utilizing photosynthate which cause accumulation of sucrose, and organic phosphate with reduction of P_i availability, in rice maternal and grain tissue, or long term down regulation of photosynthetic gene products in source tissue (e.g. Rubisco) might be alleviated by the elevation of sink strength. Increasing sink strength can drive phloem loading at the source and unloading at the seeds resulting from osmotically driven mass flow, and can prevent reloading from sink organ by sustained sucrose removal.

Feedback on photosynthesis is also observed during the vegetative phase of rice development evaluated by following transient changes in response of photosynthesis in switching from 20 to 2 kPa O_2 (Sun et al 1999a). The results indicated strong feedback inhibition at moderate temperatures (18–25 °C), atmospheric levels of CO_2 and saturating light. Again, this inhibition was interpreted as limitation on utilization of triose phosphate, which has a feedback effect on the use of reductive power, electron transport and regeneration of ribulose bisphospate (assimilatory charge). Under high temperature, feedback was only observed under high light and with CO_2 up to two times atmospheric levels. Our results indicate feedback is less likely under current atmospheric levels of CO_2 when leaf temperatures are very high, or humidity is low causing a decrease in stomatal conductance; conditions which will likely cause a CO_2 limitation and increased photorespiration (see also Murchie et al 1999, Sun et al 1999a, Winder et al 1998).

Low capacity for starch synthesis in leaves makes rice
more susceptible to feedback inhibition of photosynthesis

A limitation in triose phosphate utilization occurs when CO_2 fixation rates exceed the capacity to synthesize carbohydrates such as sucrose and starch and recycle P_i. This view is supported by carbohydrate analysis of rice leaves when plants are subjected to different temperatures. Rice leaves accumulate high levels of sucrose but very little starch when plants are grown at several different temperatures. The starch content of leaves was low at all temperatures (18, 25 and 32 °C), less than 1 mmol glucose equivalents m^{-2} (Fig. 2). The sucrose content of leaves was high, 25 mmol sucrose m^{-2}. On a fresh weight basis it is equivalent to about 175 mM,

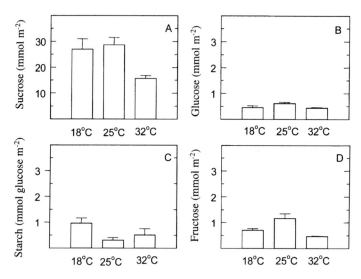

FIG. 2.   Carbohydrate contents in the leaves of rice plants grown under various temperatures at a PPFD of 700. Experiments were initiated 2 h after the start of the photoperiod under normal growth conditions (700 PPFD, 26 °C). Figure taken from Winder et al (1998).

assuming equal partitioning between the cytoplasm and vacuole. These observations suggest that, under conditions when sucrose synthesis becomes saturated, the resulting excess photosynthate cannot be converted into starch. We suggest that this inability of the rice plant to use starch as a transient sink during periods of excess photosynthate contributes to the photosynthetic feedback response. The importance of leaf starch as a transient sink and its impact on photosynthetic capacity and plant growth is readily evident in *Arabidopsis* as discussed below.

Analysis of the starch-deficient *Arabidopsis* mutants

The small size of *Arabidopsis* plants permits whole plant evaluation of photosynthetic potential using *in situ* methods. The steady-state levels of starch synthesis *in vivo* were assessed in a special chamber designed to feed $^{14}CO_2$ to whole *Arabidopsis* plants while simultaneously monitoring net CO_2 assimilation (A), true rates of photosynthetic O_2 evolution (J_{O2}, determined from chlorophyll fluorescence measurements of photosystem II), and partitioning of photosynthate into sucrose and starch, and plant growth (Sun et al 1999b). Compared to the wild-type, the starch mutants showed reduced photosynthetic capacity. The starch null mutant showed the largest decrease while the starch-deficient mutant showing intermediate levels when subjected to high light and

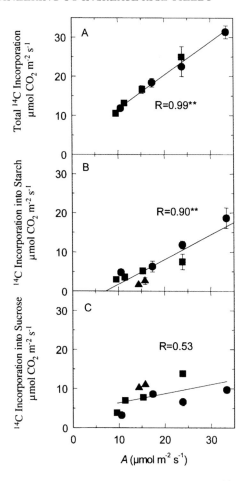

FIG. 3. The relationship between CO_2 assimilation rate, total ^{14}C incorporation, starch, sucrose synthesis in *Arabidopsis*. CO_2 assimilation was measured at a PPFD of 800 μmol m^{-2} s^{-1} at 25 °C and various CO_2 partial pressures in wild-type (closed circle), the starch-deficient mutant TL46 (closed square) and the starch null mutant TL25 (closed triangles). Different points for each genotype are the results of different CO_2 and light levels. Figure taken from Sun et al (1999b).

increased CO_2 partial pressure. Moreover, the extent of stimulation of CO_2 assimilation by increasing CO_2 or reducing O_2 partial pressure was significantly less for the starch mutants than for the wild-type with the degree inversely correlated with starch content. Overall, $^{14}CO_2$ partitioning showed that there is a strong correlation between the carbon assimilation rate and the rate of starch synthesis, but not with the rate of sucrose synthesis (Fig. 3). Moreover, this correlation could also be extended between the extent of plant growth and leaf starch levels. These results indicate that leaf starch not only serves as a transient reserve for carbon and energy during the night but also as transient sink for excess triose phosphate during the day. The conversion of triose phosphate into starch results in the recycling of P_i required for photophosphorylation and alleviates any potential feedback effects. Hence, starch synthesis is likely to play a role in effecting photosynthesis not only during the reproductive stage (grain filling) but also during the vegetative stage of rice development.

Strategies to increase rice productivity and yield by manipulation of starch

A key regulatory step in starch biosynthesis is catalyzed by the enzyme ADP glucose pyrophosphorylase (AGPase). This enzyme is allosterically regulated by the triose phosphate, 3-phosphoglyceric acid (3-PGA), which is required for catalytic activity, and by P_i, which antagonizes the activation by 3-PGA. These effector molecules are important not only in controlling AGPase activity and, in turn, the flux of carbon into starch but also regulating the rate of sucrose biosynthesis as well as photosynthesis as discussed above.

The allosteric regulatory properties of the leaf AGPase is an essential process in maintaining the partitioning of carbon between sucrose and starch. During photosynthesis, AGPase is tightly down-regulated so that the bulk of the fixed carbon is directed to sucrose biosynthesis with very little carbon entering starch (Stitt 1996). When the capacity to convert triose phosphate to sucrose is saturated, or when sucrose levels begin to buildup in the leaf cell because of limitations in sucrose export or compartmentation in the vacuole, 3-PGA and P_i levels begin to rise and fall, respectively. These conditions can result in P_i limitation and a feedback effect on photosynthesis. However, if there is capacity to synthesize starch, these conditions are favourable for activating AGPase and, in turn, increasing carbon flow into starch. These contrasting oscillatory responses in sucrose, starch and photosynthesis are important processes in maintaining plant homeostasis. They are, however, not efficient in maintaining high rates of photosynthesis to maximize plant productivity.

One approach to alleviate feedback effects is to increase the flux of carbon into leaf starch as suggested by our results obtained from a study of the *Arabidopsis* starch mutants. This change should increase P_i availability to maintain photosynthesis and at the same time have negligible effects on maintaining high rates of sucrose synthesis needed to support plant growth and development. Since AGPase plays the key regulatory role in starch biosynthesis, this enzyme would be the most appropriate target to manipulate to increase starch synthesis.

TABLE 1 Regulatory properties of various recombinant AGPases. $A_{0.5}$ is the amount of 3-PGA required to give 50% activation. $I_{0.5}$ is the amount of P_i required to inhibit the enzyme 50% in the presence of a known amount of 3-PGA

AGP Line	$A_{0.5}$ (mM)	$I_{0.5}$ (mM)	$I_{0.5}/[3\text{-}PGA]$
UpReg1	0.002	3.1	24.0
R4	0.09	1.2	4.8
R20	0.12	0.57	1.7–2.3
Wild-type	0.16	0.07	0.3–1.4
R32	0.41	0.32	1.3
M345 (P52L)	4.0	2.5	0.6

One restriction in manipulating AGPase is to maintain the allosteric regulatory properties of the enzyme so that sucrose synthesis is favoured over starch synthesis. Hence, the AGPase enzyme must be engineered so that only incremental increases in carbon flux into starch is obtained with no significant impact on sucrose synthesis. Mutant AGPase enzymes, which may exhibit these properties, have been generated and identified for the potato AGPase. By using random mutagenesis and bacterial complementation, AGPase large-subunit mutant sequences have been identified (Greene et al 1998). When expressed and assembled with the wild-type small subunit, the resulting AGPase enzyme has normal catalytic characteristics but altered allosteric regulatory properties. Table 1 compares the allosteric regulatory properties of the wild-type and several AGPase mutant enzymes. Several AGPase mutant enzymes (UpReg1, R4 and R20) have been obtained which require lower amounts of 3-PGA for catalytic activity than the wild-type enzyme. In addition, these same enzymes are more resistant to inhibition by P_i as compared to the wild-type enzyme. Based on these kinetic properties, UpReg1, R4 and R20 display up-regulatory properties and these mutant enzymes can be expected to be much more catalytically active than the wild-type enzyme. Efforts are currently being made to express these mutant AGPase genes in mature rice leaf tissue and to determine their possible impact on carbon partitioning and photosynthetic capacity during plant growth and development.

Modulation of carbon flow into seed starch

As discussed above, the flux of carbon into starch is the major process that influences the capacity of developing seeds to utilize photosynthate. A major limitation in this process is likely to be the formation of ADP glucose because of the allosteric regulatory properties of the AGPase enzyme. Although the major seed AGPases in barley and maize endosperm are either insensitive or display varying degrees of sensitivity to 3-PGA and P_i, the major rice AGPase enzyme activity requires 3-PGA for catalytic activity and is sensitive to P_i inhibition (C. Sakulsingharoj & T. W. Okita, unpublished observations). This allosteric regulating behaviour of AGP in developing seeds would have a marked effect on reducing the rate of carbon flow into starch. In the presence of 1 mM 3-PGA, the maize and rice enzymes are almost fully activated. This level of enzyme activity is reduced by 50% in the presence of only 0.44 mM P_i. In developing endosperm tissue, the actual level of net enzyme activity can be expected to be much lower as P_i levels are probably present at much higher levels than 3-PGA levels. Although the plant probably compensates for this net inhibition of AGP by P_i by increasing AGPase levels during seed development, it is likely that net carbon transfer into starch has not yet reached its maximum potential (Choi et al 1998, Nakata & Okita 1994). Manipulation of the AGPase by introducing an enzyme that no longer requires an activator molecule for catalytic activity and/or increased resistance towards P_i inhibition may be beneficial. Studies with potatoes (Giroux et al 1996, Stark et al 1992) have now demonstrated that the introduction of allosteric mutant AGPase can increase starch production resulting in increased yields.

Of the allosteric regulatory AGPase mutant enzymes available, the most catalytically active are those from bacteria. These include 618, SG5, SG14 and CL1136 (Preiss & Romeo 1994). Results from site-directed mutagenesis studies indicated that certain amino acid replacements at the original mutation site conferred even further increase sensitivity to the activator and less sensitivity to the inhibitor than the original mutation. Based on this study, modified mutations of those initially described in strains SG3, CL1136 and 618 (*glg*C16) were pyramided to yield the *glg*C triple mutant (TM). In the absence of activator, the coded TM enzyme exhibits up to 90% of the catalytic activity of the levels evident for the fully activated wild-type enzyme, an improvement of about 30% over that evident for the *glg*C16 coded enzyme (J. Preiss, personal communication).

At present, the *glg*C-TM has been transferred into a japonica rice line (Kitaake) in Pullman and the new plant type at IRRI (S. Datta, personal communication). In Pullman, transgenic plants expressing *glg*C-TM at the RNA, protein and enzyme levels have been identified. Some of these transgenic lines bear higher grain weight seeds up to 10% higher than those produced by untransformed plants. Current efforts are directed at obtaining evidence to conclusively demonstrate that increase in seed weight correlates with degree of *glg*C-TM expression and assessing the general utility of this gene in increasing grain weight in different rice genotypes.

Acknowledgements

Research described within was supported in part by DoE Grant No. DE-FG03-96ER20216, U.S. Department of Agriculture National Research Initiative Competitive Grants Program Award No. 98-35306-6469.

References

Baker JT, Allen LH, Boote KJ 1990 Growth and yield responses of rice to carbon dioxide concentration. J Agric Sci 115:313–320

Board JE, Tan Q 1995 Assimilatory capacity effects on soybean yield components and pod number. Crop Sci 35:846–851

Chen CL, Sung JM 1994 Carbohydrate metabolism enzymes in CO_2-enriched developing rice grains varying in grain size. Physiol Plant 90:79–85

Choi SB, Zhang Y, Ito H et al 1998 Increasing rice productivity by manipulation of starch biosynthesis during seed development. In: Waterlow J, Armstrong D, Fowden L, Riley R (eds) Feeding a world population of more than eight billion people: a challenge to science. Oxford University Press, New York, p 137–149

Giroux MJ, Shaw J, Barry G et al 1996 A single gene mutation that increases maize seed weight. Proc Natl Acad Sci USA 93:5824–5829

Greene TW, Kavakli IH, Kahn ML, Okita TW 1998 Generation of up-regulated allosteric variants of potato ADP-glucose pyrophosphorylase by reversion genetics. Proc Natl Acad Sci USA 95:10322–10327

Ho LC 1988 Metabolism and compartmentation of imported sugars in sink organs in relation to sink strength. Annu Rev Plant Physiol Mol Biol 39:355–378

Hocking PJ, Meyer CP 1991 Effects of CO_2 enrichment and nitrogen stress on growth and partitioning of dry matter and nitrogen in wheat and maize. Aust J Plant Physiol 18: 339–356

Imai K, Coleman D, Yanagisawa T 1985 Increase of atmospheric partial pressure of carbon dioxide and growth and yield of rice (*Oryza sativa* L.). Jpn J Crop Sci 54:413–418

Jang JC, Sheen J 1997 Sugar sensing in higher plants. Trends Plant Sci 2:208–214

Jang JC, Leon P, Zhou L, Sheen J 1997 Hexokinase as a sugar sensor in higher plants. Plant Cell 9:5–19

Koch KE 1996 Carbohydrate-modulated gene expression in plants. Annu Rev Plant Physiol Plant Mol Biol 47:509–540

Lee J-S, Daie J 1997 End-product repression of genes involving carbon metabolism in photosynthetically active leaves of sugar beets. Plant Cell Physiol 38:887–894

Leegood RC, Edwards GE 1996 Carbon metabolism and photorespiration: temperature dependence in relation to other environmental factors. In: Baker NR (ed) Photosynthesis and the environment. Kluwer Academic, Dordrecht (Adv Photosyn 5) p 191–221

Murchie EH, Chen Y-Z, Hubbart S, Peng S, Horton P 1999 Interactions between senescence and leaf orientation determine *in situ* patterns of photosynthesis and photoinhibition in field-grown rice. Plant Physiol 111:553–564

Nakata PA, Okita TW 1994 Studies to enhance starch biosynthesis by manipulation of ADP-glucose pyrophosphorylase genes. In: Belknap WR, Vayda ME, Park WD (eds) The molecular and cellular biology of the potato. CAB International, Wallingford, p 31–44

Preiss J, Romeo T 1994 Molecular biology and regulatory aspects of glycogen biosynthesis in bacteria. In: Cohn WE, Moldave K (eds) Progress in nucleic acid research and molecular biology, vol 47. Academic Press, San Diego, CA, p 299–329

Rowland-Bamford AJ, Allen LH Jr, Baker JT, Boote KJ 1990 Carbon dioxide effects on carbohydrate status and partitioning in rice. Plant Cell Environ 14:1601–1608

Rowland-Bamford AJ, Baker JT, Allen LH, Bowes G 1991 Acclimation of rice to changing atmospheric carbon dioxide concentrations. Plant Cell Environ 14: 577–583

Sage R 1990 A model describing the regulation of ribulose-1,5-bisphosphate carboxylase, electron transport, and triose phosphate use in response to light intensity and CO_2 in C3 plants. Plant Physiol 94:1728–1734

Sage RF 1994 Acclimation of photosynthesis to increasing atmospheric CO_2: the gas exchange perspective. Photosynth Res 39:351–368

Sage RF, Sharkey TD 1987 The effect of temperature on the occurrence of O_2 and CO_2-insensitive photosynthesis in field grown plants. Plant Physiol 84:658–664

Schaffer AA, Zamski E 1996 Photoassimilate distribution in plants and crops: source–sink relationships. Marcel Dekker, New York

Sharkey TD 1985 O_2-insensitive photosynthesis in C3 plants: its occurrence and a possible explanation. Plant Physiol 78:71–75

Sharkey TD 1990 Feedback limitation of photosynthesis and the physiological role of ribulose bisphosphate carboxylase carbamylation. Bot Mag Tokyo 2:87–105

Sharkey TD, Stitt M, Heineke D, Gerhardt R, Raschke K, Heldt HW 1986 Limitation of photosynthesis by carbon metabolism. Plant Physiol 81:1123–1129

Sharkey TD, Laporte MM, Micallef BJ, Shewmaker CK, Oakes JV 1995 Sucrose synthesis, temperature and plant yield. In: Mathis P (ed) Photosynthesis: from light to biosphere. Kluwer Academic, Dordrecht, p 635–640

Sheen J 1994 Feedback control of gene expression. Photosynth Res 39:427–438

Sims DA, Luo Y, Seeman JR 1998 Comparison of photosynthetic acclimation to elevated CO_2 and limited nitrogen supply in soybean. Plant Cell Environ 21:945–952

Smeekens A, Rook F 1997 Sugar sensing and sugar mediated signal transduction in plants. Plant Physiol 115:7–13

Stark DM, Timmerman KP, Barry GF, Preiss J, Kishore GM 1992 Regulation of the amount of starch in plant tissues by ADP glucose pyrophosphorylase. Science 258: 287–292

Stitt M 1986 Limitation of photosynthesis by carbon metabolism. Evidence for excess electron transport capacity in leaves carrying out photosynthesis in saturating light and CO_2. Plant Physiol 81:1115–1122

Stitt M 1996 Metabolic regulation of photosynthesis. In: Baker NR (ed) Photosynthesis and the environment. Kluwer Academic, Dordrecht (Adv Photosyn 5) p 151–191

Stitt M, Huber S, Kerr P 1987 Control of photosynthetic sucrose formation. In: Hatch MD, Boardman NK (eds) Photosynthesis, vol 10. Academic Press, San Diego, CA, p 327–409

Sun J, Okita TW, Edwards GE 1999a Feedback inhibition of photosynthesis in rice measured by O_2 dependent transients. Photosynth Res 59:187–200

Sun J, Okita TW, Edwards GE 1999b Modification of carbon partitioning, photosynthetic capacity and O_2 sensitivity in *Arabidopsis* plants which have low ADP-glucose pyrophosphorylase. Plant Physiol 119:267–276

Waterlow JC, Armstrong DG, Fowden L, Riley R 1998 Feeding a world population of more than eight billion people. Oxford University Press, New York

Winder TL, Sun J, Okita TW, Edwards GE 1998 Evidence for the occurrence of feedback inhibition of photosynthesis in rice. Plant Cell Physiol 39:813–820

Yoshida S 1981 Physiological analysis of rice yield. In: Yoshida S (ed) Fundamentals of rice crop science. International Rice Research Institute, Los Banos, Philippines

Ziska LH, Teramura AH 1992 CO_2 enhancement of growth and photosynthesis in rice (*Oryza sativa*). Plant Physiol 99:473–481

DISCUSSION

Horton: You showed some convincing evidence for this feedback inhibition phenomenon. What is not clear is how general this is. In Maurice Ku's work, the fact that he could stimulate photosynthesis by opening stomata would suggest that there is no limitation at the sink level under those conditions in that variety. The measurements that we made show that if we increase the level of CO_2 in the leaf chamber in the field, we see a huge increase in photosynthetic rate, which again tells us that it is not sink-limited (Murchie et al 1999). We have also measured carbohydrate levels in our leaves at varying periods during the day, and we don't see any significant build-up of carbohydrates. One has to be careful of extrapolating and saying that this is a fundamental feature of rice: that because rice plants do not synthesize starch in the chloroplasts this therefore means that they are feedback-inhibited. I am not doubting that you see it in your experiments, but you have to be careful of generalizing this to all situations in all varieties.

Okita: That is an excellent point. The limitation that we have is that our rice plants are grown in growth chambers. We are limited by the type of environmental conditions that the rice plants are subjected to. The overall photosynthetic capacity is dependent on the conditions the plants are grown under, and we are limited by how much light we can give them. It is something we have to be cautious about. Also, we have only looked at one rice variety, which is a japonica type.

Ku: Is anything known about the leaf AGPase in rice? Is it more sensitive to this inhibitor relative to some other species, which accumulate starch?

Okita: We see very little enzyme activity in leaf tissue, which correlates with the amount of starch seen. These low enzyme levels are also evident in our ^{14}C labelling experiments: very little of the fixed carbon actually goes into starch.

Ku: In the rice mutants that exhibit different sensitivity to P_i inhibition, do the different degrees of inhibition correlate with the amount of starch accumulated in these different mutants?

Okita: The mutant analysis has unfortunately only been done in *Escherichia coli*. The way we select for these mutants is by looking at the ability of *E. coli* cells to make glycogen. At least in *E. coli*, there is a very strong correlation between the changes in allosteric regulatory properties and glycogen biosynthesis levels. We expect to see the same correlation in plants.

Beyer: My understanding of starch biosynthesis is quite limited, but I know that it requires a whole bunch of enzymes that are organized in complexes. You consider that changing the expression of glucose pyrophosphorylase is a rate-limiting step. In chloroplasts of rice leaves are these later enzymes constitutively expressed or are they induced?

Okita: Normally the starch synthases and branching enzymes are present in most tissues. The best example I can give is of the Monsanto experiment in which they expressed the up-regulated form of AGPase in rapeseed development. This totally fouled up carbon partitioning. In this normally oil-containing plant crop they got a starch-containing organ.

Leach: What is the starch content of the roots in these rice plants?

Okita: I don't know. The major organ for starch storage in rice is the base of the stem. This is one of the factors many plant breeders use in terms of assessing the possible yield characteristics.

Leung: What would the AGPase rice mutant look like?

Okita: There are rice equivalents to the maize *shrunken 2/brittle 2* mutants, which would essentially be knockouts of the large subunit or small subunit for the cytoplasmic form. They pretty much follow the type of phenotypes you would find for the maize equivalent mutations.

Leung: Has this been reported in the rice mutant stock?

Khush: Yes, there are six brittle mutants, involving different genes.

Elliott: Perhaps it would be appropriate in a symposium such as this to say a few things that are not yet based on rock hard data. We have been doing some work in collaboration with John Bennett on exploring the extent to which the defects of grain filling of new plant type rice are a consequence of problems with either source strength, or sink strength, or both. John conceived the experiment in which we put the isopentenyl transferase (*ipt*) cytokinin biosynthesis gene under the control of the *Arabidopsis* SAG12 promoter, to see whether by delaying the senescence of the flag leaf we could improve grain filling. This has proved to be technically demanding, but we have succeeded in getting the *ipt* gene into wheat, in an active form. It delays not only senescence but also maturation. The most important point with regard to the question that we are addressing now is that the grains of the wheat are dramatically increased in size. The second thing that is important to note is that the strategy for enhancement of grain filling benefits from a two-pronged approach. First we question the extent to which grain filling is defective because the relevant grains don't have the appropriate sink capacity. We are assaulting that by a strategy that involves regulation of the *ipt* gene by endosperm-specific promoters expressed early in endosperm development. Along with cell cycle genes under the control of the same promoter, these determine an increase in cell number, which we believe will contribute to an enhancement of the sink capacity. Then the strategy rolls on to deliver enhanced auxin levels later in development under the control of an endosperm-specific promoter which is expressed later. We think that by enhancing auxin levels at this stage we will increase the sink strength.

Okita: I think these approaches are valid in terms of increasing overall sink strength. I am curious about the idea of increased auxin levels increasing sink strength: what effects will this have?

Elliott: Endosperm development in cereals is accompanied by the successive rise and fall of cytokinin, auxin and abscisic acid levels (Lur & Settler 1993, Morris et al 1993, Yuan & Huang 1993). The wave of cytokinin is associated with mitosis of endosperm cells (Dietrich et al 1995), the wave of auxin with massive accumulation of carbohydrate and protein, and the wave of abscisic acid with dehydration, ripening and dormancy.

Leach: What is the relationship between stress and starch movement?

Okita: Most of the studies that I am aware of have been done in wheat and maize, and much of the work has been done on temperature stress. Several studies have evaluated the control of starch synthesis during seed development under temperature stress. Soluble starch synthase has been suggested as the rate-limiting enzyme in starch biosynthesis when wheat plants are exposed to elevated temperature (Jenner et al 1993). In maize, it has been suggested that AGPase may be limiting under temperature stress conditions (Greene & Hannah 1998). It remains unclear, however, whether the reduction of these enzyme activities under temperature stress conditions is the causal basis for the reduction in the rate in starch biosynthesis or simply a by-product of the condition. Elevated temperatures can have a profound effect on many processes such as the transport of assimilates from the source leaves, unloading to the sink organs, and their metabolism. For example, invertase has been suggested as a limiting enzyme in the conversion of sucrose into starch during early seed embryo development (Zinselmeier et al 1999).

Gale: I'd like to ask a slightly oblique question. Whenever I hear of work on increasing starch synthesis, we always concentrate on nuclear genes. Nevertheless, starch synthesis takes place inside the plastid, and there is a plastid genome that is doing remarkable things at this time. In wheat, the amyloplast genome starts out at around 10 copies per cell at the beginning at starch filling, then there is massive endoreduplication, leading to something like 50 copies at 30 days (see Catley et al 1987). I guess the situation will be similar in rice. Nevertheless we completely ignore the plastid genome when it comes to the biochemistry of what is going on inside the grains. Are we missing anything by concentrating solely on nuclear genes?

Okita: Part of the problem is that it is only within the last five years that we have been able to transform chloroplasts. The extent of regulation of putting a transgene into chloroplasts is just not known. There are a lot of unknowns, as opposed to the expression of a transgene in the nuclear genome.

Gale: I'm not so much bothered about transgenes. Why does the amyloplast genome go through endoreduplication at this time? It must be doing something?

Mazur: This would be a good candidate for transcript analysis and proteomics.

Leach: Are there plastid genes involved in starch synthesis?

Okita: Almost all the starch synthesis genes that I am aware of are nuclear-encoded, but there will probably be plastid genes that affect starch biosynthesis.

Leung: In this work with grain filling, which mutants do you think we should be looking for in order to dissect the pathways?

Ku: That is a good question. Steve Huber at North Carolina State University has suggested that increased sucrose formation for transport to the sink tissue would help increase shoot growth. There are a few targets that can be manipulated through genetic engineering. First is fructose bisphosphatase (FBPase), which controls the key step in sucrose formation in the cytoplasm. The activity of this enzyme is regulated by the metabolite fructose-2,6-phosphate. This is an important inhibitor. Jack Preiss has done quite a bit of work on this step. The other one is sucrose phosphate synthase (SPS). Tom Sharkey has done some elegant work showing that when one SPS is over-expressed in tomato, biomass is increased by up to 30% and the crop yield increases by 40%. I heard that in Japan the Japan Tobacco group has introduced SPS into rice without seeing any substantial effect. Since these data are not yet published we don't know what is happening in these rice plants. One other possibility is to have more sucrose transport out of the source and into the sink tissues. There are a few groups working on this, but we haven't seen any positive effect. We are trying to learn the lesson from heterosis. If you do a differential display between parental corn versus hybrid corn, there are two or three major genes that are altered. One is related to fructose-2,6-phosphate metabolism: the enzyme for this synthesis is down-regulated. The other enzyme, malate dehydrogenase, is increased. This is a mitochondrial enzyme related to energy supply. We are currently checking the expression of the corresponding genes in hybrid rice. If we can suppress the first enzyme using antisense, the leaf would have less fructose-2,6-phosphate, a molecule which acts as a brake to slow down sucrose formation, thus allowing more sucrose to flow. There is still some metabolic flexibility or plasticity that would allow more sucrose to flow to the sink tissues.

Okita: I agree. In addition to mutants defective in metabolism, we should be looking for mutants defective in the overall architecture of a developing endosperm cell. The amyloplasts and protein bodies are not randomly scattered, but are instead heavily stratified in the cytoplasm. The protein bodies are along the periphery of the cell close to the plasma membrane; amyloplasts are further in. There are probably mutants in rice equivalent to the *dek* mutants in maize, which have altered cytological architecture, probably caused by cytoskeletal defects. It would be useful if we could understand in fine detail how metabolites are transported and utilized in developing sink tissues.

Horton: It is clear that in this sort of complex metabolic system we can learn a lot by looking at mutations or antisense effects. The kind of mutants I would like to get hold of are those where the photosynthetic system as a whole is not responding in the same way to environmental signals or to internal signalling processes such as sugar levels. Then we can try to understand how this fantastic balancing procedure is achieved. This is getting at the heart of what we want to understand: the plasticity in the photosynthetic system. We are trying to identify mutants that do not respond appropriately to signals. This will tell us an awful lot about the way in which this whole system fits together. In this discussion the question arose as to whether the source is limiting or whether the sink is limiting. Really, the problem is that it is quite hard to define a 'sink' or 'source' under many circumstances, and also that regulatory systems tend to try to keep these in balance. We want some way of perturbing those regulatory mechanisms in a way that will let us know how the plant 'works'. This is where the big knowledge gap is.

Dong: Howard Goodman's lab has an *Arabidopsis* mutant in which the seed accumulates starch instead of oil bodies. This is worth investigating.

Okita: This mutation is a little bit different than they think it is. In most oil-bearing seeds there is mobilization of carbon into starch first, which is then remobilized and converted to oil. The mutation may be a block in this transition process.

Khush: Several kinds of starch mutants are known in rice. For example, the dull mutant is commonly produced through mutagenesis.

Okita: I am not familiar with the rice mutant, but I am familiar with the equivalent maize mutation. This is a defect in soluble starch synthase II. What is unusual about this starch biosynthetic enzyme is that it was originally isolated and found to have a molecular mass of about 60 kDa and then later 90 kDa. But when they isolated the gene it turned out to encode a protein of 180 kDa — much larger than is needed for catalytic activity. The catalytic activity resides on a C-terminal domain about 60 kDa in size. Hence, this enzyme is very sensitive to proteolysis, giving a truncated product that still retains starch synthase activity. One question that is raised is what is the function of the long N-terminal region of the enzyme? Possibilities include that this region has another enzymatic/regulator role or that it interacts with other proteins. The latter is a distinct possibility as *dull* not only results in a loss of soluble starch synthase II but also branching enzyme IIa activity.

References

Catley MA, Bowman CM, Bayliss MW, Gale MD 1987 The pattern of amyloplast DNA accumulation during wheat endosperm development. Planta 171:416–421

Dietrich JT, Kaminek M, Bleuens DG, Rainbott TM, Morris RO 1995 Changes in cytokinins and cytokinin oxidase activity in developing maize kernels and the effects of exogenous cytokinin on kernel development. Plant Physiol Biochem 33:327–336

Greene TW, Hannah LC 1998 Enhanced stability of maize endosperm ADP-glucose pyrophosphorylase is gained through mutants that alter subunit interactions. Proc Natl Acad Sci USA 95:13342–13347

Jenner CF, Siwek K, Hawker JS 1993 The synthesis of [^{14}C]starch from [^{14}C]sucrose in isolated wheat grains is dependent on the activity of soluble starch synthase. Aust J Plant Physiol 20:329–335

Lur HS, Setter TM 1993 Role of auxin in maize endosperm development, timing of nuclear DNA endoreduplication, zein expression and cytokinin. Plant Physiol 103:273–280

Morris RO, Blevins DG, Dietrich JT et al 1993 Cytokinins in plant pathogenic bacteria and developing cereal grains. Aus J Plant Physiol 20:621–637

Murchie EH, Chen Y-Z, Hubbart S, Peng S, Horton P 1999 Interactions between senescence and leaf orientation determine in situ patterns of photosynthesis and photoinhibition in field grown rice. Plant Physiol 119:553–563

Yuan L, Huang JG 1993 Effects of potassium on the variation of plant hormones in developing seeds of hybrid rice. J Southwest Agri Univ 15:38–41

Zinselmeier C, Jeong BR, Boyer JS 1999 Starch and the control of kernel number in maize at low water potentials. Plant Physiol 121:25–36

Genetic analysis of plant disease resistance pathways

Jane E. Parker, Nicole Aarts, Mark A. Austin, Bart J. Feys, Lisa J. Moisan, Paul Muskett and Christine Rusterucci

The Sainsbury Laboratory, John Innes Centre, Norwich Research Park, Colney Lane, Norwich NR4 7UH, UK

Abstract. Plant disease resistance (R) genes are introduced into high yielding crop varieties to improve resistance to agronomically important pathogens. The R gene-encoded proteins are recognitionally specific, interacting directly or indirectly with corresponding pathogen avirulence (*avr*) determinants, and are therefore under strong diversifying selection pressure to evolve new recognition capabilities. Genetic analyses in different plant species have also revealed more broadly recruited resistance signalling genes that provide further targets for manipulation in crop improvement strategies. Understanding the processes that regulate both plant–pathogen recognition and the induction of appropriate defences should provide fresh perspectives in combating plant disease. Many recent studies have utilized the model plant, *Arabidopsis thaliana*. Here, mutational screens have identified genes that are required for R gene function and for restriction of pathogen growth in compatible plant–pathogen interactions. Genetic analyses of these plant mutants suggest that whilst signalling pathways are conditioned by particular R protein structural types they are also influenced by pathogen lifestyle. Two *Arabidopsis* defence signalling genes, *EDS1* and *PAD4*, are required for the accumulation of salicylic acid, a phenolic molecule required for systemic immunity. The cloning, molecular and biochemical characterization of these components suggests processes that may be important in their disease resistance signalling roles.

2001 Rice biotechnology: improving yield, stress tolerance and grain quality. Wiley, Chichester (Novartis Foundation Symposium 236) p 153–164

In the natural environment, plants encounter many different types of pathogen and have evolved mechanisms to perceive and respond effectively to attack. Survival of the plant relies on the early recognition of pathogen-derived molecules and the timely activation of local and systemic defences (Parker 2000). Although plants lack a circulation system that exists in multicellular animals for the surveillance and removal of foreign material, they possess an operationally equivalent recognition system in the form of a basal resistance machinery and multiple recognitionally specific Resistance (R) genes (Martin 1999, McDowell & Dangl 2000). Understanding the molecular basis of plant–pathogen recognition and the

153

interplay of plant defence signalling pathways is now a major goal in plant pathology and should offer new perspectives in combating some of the world's most destructive diseases of crop plants.

The expression of dominant or co-dominant R genes is readily scorable in plant crosses or in transgenic plant lines and therefore these traits are utilized extensively in crop protection programmes against major pathogens (Pink & Puddephat 1999). This type of resistance is commonly, although not invariably, associated with localized plant cell necrosis (the hypersensitive response, HR) that, in turn, leads to the generation of systemic signals immunizing the plant against subsequent pathogen attack (known as systemic acquired resistance, SAR) (McDowell & Dangl 2000, Parker 2000). The very specificity of R gene-mediated plant responses against defined pathogen avirulence (*avr*) gene products has permitted a rigorous examination of pathogen-induced plant defences at the molecular and biochemical levels. Various studies implicate ion fluxes, protein phosphorylation and the generation of reactive oxygen intermediates (ROI) and nitric oxide as important early defence processes (McDowell & Dangl 2000, Parker 2000). Roles for the signalling molecules, salicylic acid (SA) and jasmonic acid (JA) have also been identified both in R gene-mediated responses and in systemic disease resistance pathways (Glazebrook 1999, Parker 2000). This chapter will describe some recent advances in the genetic dissection of plant disease resistance pathways and examination of the roles of several important regulatory proteins that are essential for the proper decoding of pathogen signals and initiation of plant defence.

Unravelling *R* gene-specified resistance

In recent years, molecular genetic analysis has provided one of the most powerful tools to unravel plant disease resistance pathways through the isolation of the R genes themselves from dicotyledonous and monocotyledonous plant species and the characterization of other genes that are required for R gene function (Parker 2000). Mutational screens have also identified plant loci that contribute to overall restriction of pathogen growth in compatible plant–pathogen interactions (Glazebrook et al 1996, Parker et al 1996) or that play a role in negative regulation of plant resistance pathways (Morel & Dangl 1997). The latter class of genes most probably function to 'dampen' defence responses, preventing uncontrollable and energetically costly reactions.

The most extensive mutational analyses have utilized the model crucifer, *Arabidopsis thaliana* as a tractable plant for genetics and gene cloning. Its relatively compact (~ 130 Mb) genome sequence is now complete, creating a highly informative data system to unravel plant signalling networks. Importantly, *Arabidopsis* is a host to the major plant pathogen classes: viruses, bacteria, fungi, and nematodes, and genetic variation exists in all of these interactions. Moreover, R genes cloned in *Arabidopsis* are structurally related to those isolated from crop plants. We therefore anticipate that fundamental mechanisms underlying plant–pathogen recognition and signalling in disease resistance will be broadly conserved across plant taxonomic boundaries. What emerges from various analyses in *Arabidopsis* is a potentially highly branched informational network with multiple signal amplification loops (Glazebrook 1999, McDowell & Dangl 2000, Parker 2000). Superimposed on these signalling circuits are processes that establish hierarchies of pathway utilization. These, in turn, are influenced both by R protein-mediated recognition events and by the lifestyle of the pathogen invading the plant. The challenge now is to identify the key regulatory components in these pathways and comprehend how they function within the plant cell.

Components of *RPP5*-mediated resistance in *Arabidopsis*

In our group, we have used molecular genetics in *Arabidopsis* to identify components of R gene-mediated resistance to the obligate oomycete pathogen, *Peronospora parasitica* (Parker et al 2000). The interaction between these two partners represents a highly co-evolved, natural plant–pathogen system and this is reflected in the large number (>20) of recognitionally specific *RPP* (Resistance to *P. p*arasitica) loci identified and mapped and the corresponding genetic diversity of *P. parasitica* isolates (Holub & Beynon 1997). *Arabidopsis RPP* genes isolated so far are members of the most prevalent structural class of R genes that encodes putatively cytoplasmic proteins possessing a central nucleotide-binding (NB) motif and C-terminal leucine-rich repeats (LRRs). These so called 'NB-LRR' type R proteins have been further categorized into two subclasses based on the presence of different N-termini. Class I (TIR-NB-LRR) proteins that include *RPP5* (Parker et al 1997) and members of the *RPP1* complex resistance locus (Botella et al 1998), have sequence similarities to the cytoplasmic domains of animal Toll-like receptors functioning in innate immunity. Class II (LZ-NB-LRR) proteins, of which *RPP8* (McDowell et al 1998) and *RPP13* (Bittner-Eddy et al 2000) are representatives, possess an N-terminal putative leucine-zipper (LZ) protein interaction domain. Elucidating the precise molecular events that underpin NB-LRR type R protein function is now an area of intense study, some clues possibly arising from similarities in their domain architecture with animal adaptor proteins that mediate programmed cell death (Van der Biezen & Jones 1998, Parker 2000).

Mutational screens for suppressors of *RPP5*-mediated disease resistance revealed several genes that are essential components of the plant resistance response, besides identifying multiple defective alleles of the *RPP5* gene itself

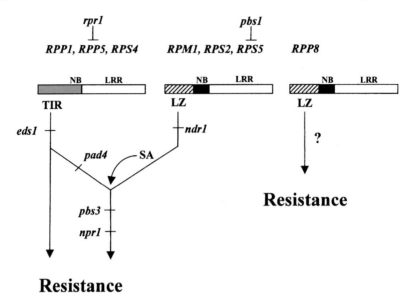

FIG. 1. In *Arabidopsis*, the characterization of mutants affecting disease resistance has permitted a genetic dissection of defence signalling mechanisms. At least three distinct R gene-mediated pathways have been defined that are influenced by the predicted R protein structural type. Whilst EDS1 is an essential component of resistance specified by TIR-NB-LRR type R proteins, NDR1 is recruited by most LZ-NB-LRR type proteins. An exception to this trend is the *RPP8* gene, encoding a LZ-NB-LRR protein that has no strong requirement for either EDS1 or NDR1. Cross-talk exists between these pathways and pathway utilization may be strongly influenced by the type of pathogen invading the plant. Both EDS1 and PAD4 lie upstream of SA-mediated defences. Perception of SA is, at least in part, mediated by the *NPR1* gene. Mutations, such as *pbs3*, define points of convergence in plant defence, although their position relative to SA is unclear. Other mutations, such as *rpr1* and *pbs1*, appear to affect specific R genes, adding further complexity to these envisaged plant defence signalling networks (see text for more details).

(Parker et al 2000). The most extensively characterized of these 'ancillary' components is the *EDS1* gene (Parker et al 1996, Falk et al 1999). Null mutations of *EDS1* abolish resistance specified by *RPP5* and several other *RPP* loci, as well as *RPS4*-mediated resistance to a bacterial pathogen, *Pseudomonas syringae* expressing the corresponding *avr* gene, *avrRps4*. Resistance can be rescued in *eds1* plants by application of a functional analogue of SA. This feature, together with analysis of defence-related gene induction profiles in wild-type and *eds1* plants has established that *EDS1* functions upstream of SA-inducible defences (Parker et al 1996, Falk et al 1999). Therefore, the wild-type EDS1 protein acts in a convergent signalling pathway conditioned by different R genes recognizing multiple pathogen types (Fig. 1). We found, however, that *EDS1* is not strongly required for resistance specified by several other *Arabidopsis* R genes, that instead recruit a different signalling component, *NDR1*, to elicit resistance (Century et al 1995, Aarts et al 1998). Interestingly, the particular signalling mode appears to be correlated with the predicted R protein structural type. Thus, the TIR-NB-LRR type R proteins isolated so far fully engage EDS1, whereas the majority of LZ-NB-LRR type R proteins are strongly dependent on NDR1. There is at least one exception to this scheme: the *RPP8* gene encodes a LZ-NB-LRR type R protein that requires neither EDS1 nor NDR1, and therefore defines a different signalling pathway (Fig. 1) (Aarts et al 1998, McDowell et al 2000). Altogether, the data suggest the existence of certain distinct downstream processes that are governed by particular R protein types, possibly activated through different N-terminal 'effector' domains. However, this linear representation of R protein signalling is undoubtedly too simplistic and more detailed characterization of these and other defence signalling mutations reveals additional points of pathway divergence, convergence, and signal cross-talk, highlighting the existence of a rather complex signalling network (Glazebrook 1999, McDowell & Dangl 2000, Parker 2000). For example, the *rpr2* mutation identified in our screens (M. A. Austin, P. Muskett & J. E. Parker, unpublished data) and *pbs3*, a suppressor of *RPS5*-mediated resistance (Warren et al 1999) are required for the function of both TIR- and LZ-NB-LRR structural R protein classes (Fig. 1). Significantly, two *Arabidopsis* mutations, *pbs1* (Warren et al 1999) and *rpr1*(M. A. Austin & J. E. Parker, unpublished data) appear to affect specific R genes, respectively, *RPS5* and *RPP5* (Fig. 1). These characteristics suggest the presence of highly discriminatory molecules that may interact with the R-Avr protein in a plant–pathogen recognition complex.

In our mutant screens, we also identified a defective allele of *PAD4* (Parker et al 2000), a gene that was originally isolated in a pathology-based screen for increased susceptibility to a virulent isolate of *P. syringae* (Glazebrook et al 1996). Phenotypic characterization of *pad4* plants has shown that wild-type *PAD4* encodes a regulatory component of phytoalexin (camalexin) production and SA accumulation after *P. syringae* inoculation (Zhou et al 1998). Significantly, *pad4* partially compromises *RPP5*-specified resistance to *P. parasitica* in contrast to *eds1* that causes a complete loss of *RPP5* function. The partial resistance phenotype of *pad4* is accompanied by an ineffective HR and the pathogen mycelium is able to grow some way beyond the point of infection (J. E. Parker, unpublished data). Thus, in a model of R gene-mediated signalling pathways (Fig. 1), we have placed *PAD4* on a bifurcation point downstream of *EDS1*-regulated processes and upstream of SA accumulation.

In addition to suppressing R gene-conditioned responses, an interesting phenotype of both *eds1* and *pad4* plants is enhanced disease susceptibility ('eds') in several compatible or partially compatible *Arabidopsis*–pathogen interactions

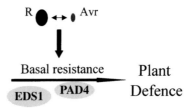

FIG. 2.   The plant defence genes, *EDS1* and *PAD4*, are essential components of R gene-mediated resistance but also contribute to the restriction of growth of several virulent pathogens. This suggests a model in which the EDS1 and PAD4 proteins operate in a basal resistance mechanism that is engaged or reinforced by particular R–Avr protein recognition events.

(Glazebrook et al 1996, Aarts et al 1998, Reuber et al 1998). This is of particular interest to us since it suggests that there are common features between processes involved in *R-avr* gene-mediated resistance (normally associated with the HR) and the less well defined restriction of virulent pathogens in disease. The data are consistent with the notion that EDS1 and PAD4 may function as general or 'basal' plant resistance regulators that have been recruited by particular types of R proteins in *R-avr* gene-stimulated responses (Fig. 2). This may become a molecular hallmark a number of plant disease resistance signalling components.

Insights to the biochemical functions of EDS1 and PAD4

The *Arabidopsis EDS1* (Falk et al 1999) and *PAD4* (Jirage et al 1999) genes were cloned and shown to encode putatively cytoplasmic proteins of 72 kDa and 60 kDa, respectively. Interestingly, although neither predicted protein exhibits extensive sequence homology to other proteins in the databases, both EDS1 and PAD4 possess within their N-terminal portions discrete motifs that resemble the catalytic sites of eukaryotic lipases (Falk et al 1999, Jirage et al 1999). The regions of amino acid similarity span three critical residues: a serine, an aspartic acid and a histidine that form a catalytic triad in a wide range of serine esterase enzymes (Schrag & Cygler 1997). The broader consensus of amino acids across these segments between EDS1, PAD4 and L-family neutral lipases, such as triacylglycerol lipases from *Rhizomucor miehei* or *R. niveus* (Schrag & Cygler 1997), suggests that the disease resistance signalling roles of EDS1 and PAD4 may be through hydrolysis of a lipid-based substrate. This may create an active fatty acid molecule, which then elicits the next step in the disease resistance pathway, or possibly degrades a negatively acting component, leading to activation of the resistance response. It is perhaps significant that several secreted and intracellular forms of mammalian platelet activating factor (PAF) acetylhydrolases that have phospholipase A2 activity, also possess a neutral lipase signature (Stafforini et al 1997). One of the functions of PAF acetylhydrolases is to process phospholipid molecules that have been fragmented by oxidation during inflammation. Since the rapid generation of ROI is a crucial early event in several *R–avr* gene-associated plant responses (McDowell & Dangl 2000, Parker 2000), it is conceivable that EDS1 and PAD4 may hydrolyze an oxygenated lipid signal. In this context, oxygenated plant lipids such as JA have been shown to play a role in the plant wound response and in certain plant–pathogen interactions (Reymond & Farmer 1998). Also, increasingly refined biochemical analysis of plant lipid metabolites suggests broader signalling attributes in these molecules than previously realized (Farmer et al 1998). One important aspect of our future studies will be to establish the precise biochemical functions of EDS1 and PAD4 in the plant defence response, both at the level of R gene-mediated defence and in 'basal' plant disease resistance. Interestingly, although both gene products are placed upstream of SA accumulation, SA up-regulates the expression of EDS1 and PAD4 as part of a signal amplification system (Falk et al 1999, Jirage et al 1999). Another component of our future work will be to isolate homologues of *EDS1* and *PAD4* from commercially important crop plants such as rice, wheat and barley, as a first step to examine their structure and function in crop systems.

Concluding remarks

Genetic analyses of plant disease resistance pathways, most extensively in the model plant, *Arabidopsis*, have provided some first insights to the complexities of signalling processes in plant–pathogen interactions. The existence of a signal network with multiple junctions for signal amplification and signal quenching probably allows the plant to engage mechanisms that are most effective against a particular pathogen and suppress inappropriate defences. It is now clear that whilst resistance pathway utilization is, in part, determined by the structural type of R protein employed in recognition, it is also strongly influenced by the pathogen's mode of attack and colonization. Thus, in designing innovative strategies to protect crop plants from pathogens and move agriculture away from its strong reliance on agrochemicals we need to comprehend the key regulatory processes that orchestrate plant stress responses. In this way, it is envisaged that we may enhance plant resistance to the most destructive pathogens without making them vulnerable to a different pathogens or insect pests, or placing too high an energy cost on plant metabolism. This is indeed a challenge and the complementary scientific approaches of plant genetics, genomics, biochemistry and transgenic technologies, now offer a real opportunity to improve plant productivity.

Acknowledgements

Research at The Sainsbury Laboratory is supported by The Gatsby Charitable Foundation and grants from The British Biotechnology and Biological Sciences Research Council and the European Commission.

References

Aarts N, Metz M, Holub E, Staskawicz BJ, Daniels MJ, Parker JE 1998 Different requirements for *EDS1* and *NDR1* by disease resistance genes define at least two R gene-mediated signaling pathways in *Arabidopsis*. Proc Natl Acad Sci USA 95:10306–10311

Bittner-Eddy PD, Crute IR, Holub EB, Beynon JL 2000 *RPP13* is a simple locus in *Arabidopsis thaliana* for alleles that specific downy mildew resistance to different avirulence determinants in *Peronospora parasitica*. Plant J 21:177–188

Botella MA, Parker JE, Frost LN et al 1998 Three genes of the Arabidopsis *RPP1* complex resistance locus recognize distinct *Peronospora parasitica* avirulence determinants. Plant Cell 10:1847–1860

Century KS, Holub EB, Staskawicz BJ 1995 *NDR1*, a locus of *Arabidopsis thaliana* that is required for disease resistance to both a bacterial and a fungal pathogen. Proc Natl Acad Sci USA 92:6597–6601

Falk A, Feys BJ, Frost LN, Jones JDG, Daniels MJ, Parker JE 1999 *EDS1*, an essential component of R gene-mediated disease resistance in *Arabidopsis* has homology to eukaryotic lipases. Proc Natl Acad Sci USA 96:3292–3297

Farmer E, Weber H, Vollenweider S 1998 Fatty acid signaling in *Arabidopsis*. Planta 206:167–174

Glazebrook J 1999 Genes controlling expression of defense responses in *Arabidopsis*. Curr Opin Plant Biol 2:280–286

Glazebrook J, Rogers EE, Ausubel FM 1996 Isolation of *Arabidopsis* mutants with enhanced disease susceptibility by direct screening. Genetics 143:973–982

Holub EB, Beynon JL 1997 Symbiology of mouse-ear cress (*Arabidopsis thaliana*) and oomycetes. Adv Bot Res 24:227–273

Jirage D, Tootle TL, Reuber TL et al 1999 *Arabidopsis thaliana PAD4* encodes a lipase-like gene that is important for salicylic acid signaling. Proc Natl Acad Sci USA 96:13583–13588

Martin GB 1999 Functional analysis of plant disease resistance genes and their downstream effectors. Curr Opin Plant Biol 2:273–279

McDowell JM, Dangl JL 2000 Signal transduction in the plant immune response. Trends Biochem Sci 25:79–82

McDowell JM, Dhandaydham M, Long TA et al 1998 Intragenic recombination and diversifying selection contribute to the evolution of downy mildew resistance at the *RPP8* locus of Arabidopsis. Plant Cell 10:1861–1874

McDowell JM, Dangl JL, Holub EB 2000 Downy mildew (*Peronospora parasitica*) resistance genes in Arabidopsis vary in functional requirements for *NDR1*, *EDS1*, *NPR1*, and salicylic acid accumulation. Plant J 22:523–529

Morel JB, Dangl JL 1997 The hypersensitive response and the induction of cell death in plants. Cell Death Differ 4:671–683

Parker JE 2000 Signalling in plant disease resistance. In: Beynon JL, Dickenson M (eds) Molecular plant pathology. Sheffield Academic Press, Sheffield (Annu Plant Rev vol 4) p 144–174

Parker JE, Holub EB, Frost LN, Falk A, Gunn ND, Daniels MJ 1996 Characterization of *eds1*, a mutation in Arabidopsis suppressing resistance to *Peronospora parasitica* specified by several different *RPP* genes. Plant Cell 8:2033–2046

Parker JE, Coleman MJ, Szabò V et al 1997 The Arabidopsis downy mildew resistance gene *RPP5* shares similarity to the Toll and Interleukin-1 receptors with *N* and *L6*. Plant Cell 9:879–894

Parker JE, Feys BJ, Van der Biezen EA et al 2000 Unravelling *R* gene-mediated disease resistance pathways in *Arabidopsis*. Mol Plant Pathol 1:17–24

Pink D, Puddephat I 1999 Deployment of disease resistance genes by plant transformation — a 'mix and match' approach. Trends Plant Sci 4:71–75

Reuber TL, Plotnikova JM, Dewdney J, Rogers EE, Wood W, Ausubel FM 1998 Correlation of defence gene induction defects with powdery mildew susceptibility in *Arabidopsis* enhanced disease susceptibility mutants. Plant J 16:473–485

Reymond P, Farmer EE 1998 Jasmonate and salicylate as global signals for defense gene expression. Curr Opin Plant Biol 1:404–411

Schrag JD, Cyglar M 1997 Lipases and alpha/beta hydrolase fold. Methods Enzymol 284:85–107

Stafforini DM, McIntyre TM, Zimmerman GA, Prescott SM 1997 Platelet-activating factor acetylhydrolases. J Biol Chem 272:17895–17898

Van der Biezen EA, Jones JDG 1998 The NB-ARC domain: a novel signalling motif shared by plant resistance gene products and regulators of cell death in animals. Curr Biol 8:R226–R227

Warren RF, Merritt PM, Holub E, Innes RW 1999 Identification of three putative signal transduction genes involved in *R* gene-specified disease resistance in *Arabidopsis*. Genetics 152:401–412

Zhou N, Tootle TL, Tsui, F, Klessig, DF, Glazebrook J 1998 *PAD4* functions upstream from salicylic acid to control defense responses in Arabidopsis. Plant Cell 10:1021–1030

DISCUSSION

Leach: Can you describe the phenotypes of *eds1* and *pad4* mutants in greater detail? Did you say that they were lesion mimics?

Parker: The *eds1* and *pad4* mutations are not lesion mimics. They suppress the resistance response to differing degrees. *Lsd1* is a lesion mimic; the wild-type gene encodes a probable dampening regulator of plant cell death. This has been defined by Jeff Dangl's characterization of the mutation. In the mutant line there is characteristic runaway cell death. If this mutation is placed in combination with *eds1*, the lesion mimic phenotype is completely suppressed. Runaway cell death induced in *lsd1* by a number of different factors is somehow feeding back into *EDS1*. This may reinforce the idea that *EDS1* has a pivotal function in a sort of basal resistance mechanism.

G.-L. Wang: Have you made double mutants of *pad4* and *eds1*?

Parker: Yes, we have constructed this double mutant, and it has an *eds1* phenotype. This reveals that the *eds1* phenotype is predominant, so it would place *pad4* downstream of *eds1*. But the compatibility conferred by the *eds1* mutation may be the absolute level of compatibility that can be attained in the plant. The genetics certainly suggest that *PAD4* is either downstream or coincident with *EDS1*.

Dong: Have you tried to look at the callose deposition in the double mutant?

Parker: No, but there is no necrosis: callose is normally associated with this.

Dong: Have you tried to overexpress wild-type *EDS1*?

Parker: We are doing that now. The plants are resistant, but we don't know whether they are hyper-resistant, and we don't know what the breadth of their resistance is to compatible pathogens.

Leach: Are there any genes in similar pathways to those you have just described found in monocots?

Parker: We have just started on this. We have found sequence homologies with monocot-derived ESTs, for example. Now we are sequencing out from these ESTs to establish whether they are *EDS1* or *PAD4* homologues. We have identified a class of lipase-like genes that are certainly present in monocots.

Salmeron: Paul Schulze-Lefert (see Pifanelli et al 1999) has mutations in genes he calls *Rar*, which are required for barley powdery mildew resistance.

Parker: Yes these are *rar1* and *rar2* mutations. It would be very interesting to see where these fit in with respect to *EDS1* in the signalling system.

Leach: Also, I would like to bring your attention to some work on phospholipase D by Sam Wang. He has antisense of different phospholipase Ds in *Arabidopsis* and is seeing some differences in interactions with pathogens. You targeted the phospholipase A in the animal system pathway, but other phospholipases could play a role. It may not be the JA pathway that is important: perhaps you want to look at other phospholipases or other lipase-like proteins.

Parker: That is a good point. I mentioned the phospholipase A2 type proteins because the homology that we identify is actually found in a subset of these phospholipase A2s, but it is also common in neutral lipases. This is completely uncharted ground. A number of groups are beginning to identify phospholipase function in plant defence signalling.

Salmeron: I have a question concerning the molecular similarities between the R gene pathway signalling and the mammalian inflammatory response. There have been some recent papers describing the bacterial lipoproteins that are commonly activators of the mammalian pathway. Lipid is an essential component of this. Have you thought about *EDS1* actually having a role in generating a signal?

Parker: Yes, we have thought about it. Actually, we have done site-directed mutagenesis of the predicted lipase catalytic amino acids that we defined by sequence in *EDS1*. We put these mutants back into the *eds1* mutant background and found that these mutations do not compromise resistance. Our first interpretation of this is that the lipase motif is not required for the resistance. Perhaps these are lipases, but this is not part of their resistance function. Alternatively, they are not lipases. However, since we have shown in a yeast two-hybrid analysis that *EDS1* can interact with *PAD4*, there is a formal possibility that perhaps *PAD4* is complementing the lipase function that *EDS1* has lost. What we have done, therefore, is put the site-direct mutants into a double *eds1*/ *pad4* mutant background, but we don't have the data yet. I think we have to be careful here; it might be that this lipase motif is a red herring. If this is so, why has it been recruited by these two signalling proteins that operate in a related mechanism?

G.-L. Wang: Is *PAD4* also required for the resistance genes with TIR/NBS/LRR motifs?

Parker: Within the scope of the analysis we have carried out *pad4* seems to affect the same resistance gene responses as *eds1*. However, the phenotypes are very different. *PAD4* is a component of the consolidation of resistance. Its utilization becomes blurred somewhat in the Rps2-mediated resistance. There we see a relaxation of resistance in *pad4*. If you look closely at the *eds1* mutation, this also slightly relaxes the Rps2-mediated resistance, so it is following the same trend. However, there is definite pathway cross-talk.

Dong: When you introduce the idea of basal resistance, I think you can explain all the phenotypes of *eds1* and *pad4* by the failure of production of SA and reactive oxygen species. Both HR and SAR require SA.

Parker: I don't think that is entirely the case. We know that if the plant fails to accumulate SA, it does not abolish the HR in the *P. parasitica* interaction. In this plant–pathogen interaction at least, the generation of SA is part of the resistance, but the other component might be ROI or something else. ROI are a marker for that particular pathway, but whether they are involved in resistance is another question.

Bennett: Is there any evidence in plants that cytochrome C is released from mitochondria as part of the programmed cell death mechanism?

Parker: Not as far as I know. There is a lot of debate about whether programmed cell death as understood in animal systems also exists in plants. I doubt that the details will be the same. There has been no identification of caspase-type protein sequences, even though caspase-like activities have been claimed to be present in plants. Also, there are a number of R gene-mediated resistances that are mediated by the same classes of R proteins that I have described that do not have an HR but they are likely to operate by a similar mechanism. I think it is doubtful that we are dealing with the same kind of programmed cell death that occurs in animals.

Beyer: How are the ROI generated if it is not by the respiratory burst machinery?

Parker: There is no doubt about the generation of ROI in the plant–pathogen interaction. However, the origin of those ROI is unclear. Although homologues to at least part of the NADPH oxidase complex, known to be the key mechanism to generate superoxide in mammalian macrophages, exist in plants, it has not been shown that they have a role to play in the generation of ROI. But the fact that ROI are produced and perceived, is clear.

Bennett: On that basis, is the difference between the animal and plant systems the predominance of peroxidases in plants?

Parker: It is clear that plants have to contend with different redox situations than animals, because of photosynthesis and its associated metabolism. You alluded to the peroxidases: there is evidence from the work, for example, of Paul Bolwell (1999) suggesting that peroxidases are a good source of ROI in the plant apoplast.

References

Bolwell GP 1999 Role of active oxygen species and NO in plant defence responses. Curr Opin Plant Biol 2:287–294

Piffanelli P, Devoto A, Schulze-Lefert P 1999 Defence signalling pathways in cereals. Curr Opin Plant Biol 2:295–300

Regulation of systemic acquired resistance by NPR1 and its partners

Xinnian Dong, Xin Li, Yuelin Zhang, Weihua Fan, Mark Kinkema and Joseph Clarke

DCMB Group, LSRC Building, Research Drive, Duke University, Durham, NC 27708-1000, USA

Abstract. The NPR1 protein of *Arabidopsis thaliana* has been shown to be an important regulatory component of systemic acquired resistance (SAR). Mutations in the *NPR1* gene block the induction of SAR by the signal molecule salicylic acid (SA). NPR1 contains an ankyrin repeats and a BTB domain which are involved in interaction with other protein(s). To further study the function of NPR1 and the regulatory mechanism of SAR, we used both molecular and genetic approaches to identify additional SAR regulatory components. Through a yeast two-hybrid screen we found that NPR1 interacts specifically with bZIP transcription factors. The involvement of bZIP transcription factors in controlling the SA-induced genes had been suggested by a number of promoter studies performed on these genes. It was found that *as1* element, which is a binding site for bZIP transcription factors, is essential for SA-induced gene expression. In a genetic screen for suppressors of *npr1*, we found a mutant, *sni1*, that restored the responsiveness to SAR induction in *npr1*. The genetic characteristics of the *sni1* mutant and the sequence of SNI1 suggest that the wild-type SNI1 protein is a negative regulator of SAR. We believe that SAR is controlled by both positive regulators and negative regulators.

2001 Rice biotechnology: improving yield, stress tolerance and grain quality. Wiley, Chichester (Novartis Foundation Symposium 236) p 165–175

The long-term goal of the research in my laboratory is to use *Arabidopsis thaliana* as a model system to determine the signalling events that lead to plant resistance to pathogens. Two plant resistance responses have been studied at the molecular level, the hypersensitive response (HR) and systemic acquired resistance (SAR).

HR is a rapid plant response to pathogens carrying an avirulence (*avr*) signal that is recognized by the host through a receptor encoded by the resistance (R) gene (Flor 1971, Hammond-Kosack & Jones 1996). The *avr*–R interaction triggers a series of physiological changes in the infected cell, which include a burst in formation of reactive oxygen species, transient Ca^{2+} fluxes, accumulation of salicylic acid (SA) and programmed cell death, which act in concert to restrict the growth of pathogens at the site of infection. Recently, many R genes have been cloned from different plant species. These R gene products are responsible for conferring resistance to pathogens ranging from viruses, bacteria and fungi, to nematodes (Bent 1996, Staskawicz et al 1995). Intriguingly, most of these R gene products seem to share conserved domains such as the leucine-rich repeat (LRR) domain, implicating the existence a common molecular mechanism in triggering the HR (Bent 1996, Staskawicz et al 1995). The significance of R-gene-triggered resistance in plants has been underscored by the discovery of large R gene clusters in all the plants surveyed (Hammond-Kosack & Jones 1997). Genome sequencing in *Arabidopsis* has indicated that ~5% of its genome encode R gene products.

SAR is a secondary resistance response that can be established after a local HR (Kuc 1982, Ross 1961, Ryals et al 1996). Once established, SAR can last from days to weeks. Unlike the immune response in animals, SAR is a broad-spectrum and non-specific resistance; local infection by a viral pathogen can trigger a systemic resistance to viruses as well as bacteria and fungi, which are otherwise virulent on the host. At present, SAR is believed to be conferred by induction of many downstream genes known as pathogenesis-related (PR) genes. In Arabidopsis, three PR genes, *PR1*, *BGL2* (also called *PR2*) and *PR5* have been cloned and used as markers for the onset of SAR (Uknes et al 1992).

It has been found that SA is a signal necessary and sufficient for eliciting SAR. During the HR, SA synthesis increases dramatically to micromolar concentrations in infected local tissues and to nanomolar concentrations systemically (Enyedi et al 1992, Meuwly et al 1995). Exogenous application of SA or SA analogues such as 2,6-dichloroisonicotinic acid (INA), induces SAR (White 1979, Métraux et al 1991) and removal of SA by the SA-degrading enzyme salicylate hydroxylase (encoded by the bacterial *nahG* gene) inhibits the induction of SAR (Gaffeny et al 1993).

Results

Multiple genetic screens have been carried out in my laboratory to identify mutants affecting SAR. In these screens, we have used the *BGL2–GUS* reporter gene whose expression is responsive to SAR induction. Transgenic lines containing this reporter were mutagenized and possible SAR-related mutants were identified for their aberrant reporter expression patterns. The *npr1-1* mutant was found for its lack of GUS activity in the presence of SA or INA (Cao et al 1994). In *npr1-1*, SA-induced expression of all three endogenous PR genes is also blocked and so is SA-induced resistance. Because *npr1-1* is recessive, we believe that the wild-type NPR1 is a positive regulator of PR gene expression and SAR. Interestingly, different screening strategies, which led to the identification of multiple *npr1* alleles (total of 12) (Cao et al 1994, 1997, Delaney et al 1995, Glazebrook et al

1996, Ryals et al 1997, Shah et al 1997), found no other genetic loci that inhibit SA induction of SAR.

The *NPR1* gene of *Arabidopsis* was cloned in my laboratory (Cao et al 1997) and its essential role in regulating SAR, as suggested by genetic evidence, was demonstrated molecularly (Cao et al 1998). Exogenous application of SA or INA results in a moderate increase in the endogenous NPR1 protein. This increase can also be achieved by expressing *NPR1* cDNA under control of the constitutive 35S promoter. The NPR1-overexpressing transgenic plants (*35S::NPR1*) exhibited the same degree of protection against pathogen infection as plants treated with exogenous SA or INA. When infected with a bacterial pathogen *Pseudomonas syringae* pv *maculicola* ES4326 (*P.s.m.* ES4326), which causes leaf-spot on wild-type *Arabidopsis*, a 1000-fold reduction in growth of this pathogen was detected in the NPR1-overexpressing plants compared to wild-type. This increase in NPR1 protein levels also led to complete resistance to an oomycete pathogen *Peronospora parasitica* Noco2 (*P. parasitica* Noco2) which normally forms downy mildew on wild-type plants (Cao et al 1998).

How did overexpression of NPR1 bring about enhanced resistance? An examination of *PR* gene expression pattern showed that overexpression of NPR1 did not result in constitutive *PR* gene expression in the absence of SA or pathogen challenge. This indicates that the NPR1 protein requires activation to be functional. In the presence of SA, INA or pathogens, a more dramatic *PR* gene induction was observed in the NPR1-overexpressing plants than wild-type, suggesting that this concerted increase in expression of *PR* genes might be the cause of the heightened resistance. There was a positive correlation between the levels of NPR1 and the enhanced induction of *PR* genes and resistance (Cao et al 1998).

The function of NPR1 is conserved in plant species. This is demonstrated by the presence of NPR1 homologues in other plants, such as tomato, tobacco, and rice, and by the ability of *Arabidopsis* NPR1 to function in a heterologous background. Overexpression of *Arabidopsis* NPR1 in rice resulted in enhanced resistance to blight caused by *Xanthomonas oryzae* (P. Ronald, personal communication). We have yet to determine the full resistance-spectrum provided by NPR1 in different plant species including *Arabidopsis*.

A number of NPR1 homologues were also found in *Arabidopsis* through the international *Arabidopsis* genome-sequencing project. Compiling sequences of NPR1 and its homologues revealed a high degree of conservation in specific protein domains (Fig. 1).

The NPR1 protein contains at least two domains involved in protein–protein interactions, the BTB/POZ (Broad-Complex, Tramtrack, and Bric à brac/ *pox*virus and *z*inc finger) domain and the ankyrin-repeat domain. In the C terminal region, a putative nuclear localization sequence (NLS) was discovered (Fig. 1) suggesting that NPR1 may function in the nucleus.

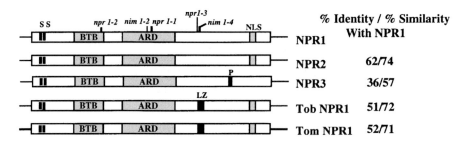

FIG. 1.   Alignment of NPR1 and its homologues. A drawing showing conserved domains between *Arabidopsis* NPR1 and NPR1 homologues (*Arabidopsis* NPR2, NPR3, tobacco NPR1 and tomato NPR1). ARD, ankyrin repeat domain; BTB, BTB/POZ domain; LZ, putative leucine zipper; NLS, nuclear localization sequence; P, ATP/GTP binding site motif A (P-loop); SS, putative phosphorylation sites.

The subcellular localization of NPR1 has been studied using an NPR1–GFP (green fluorescent protein) fusion. Transformation of this fusion gene into *npr1* rescued all the mutant phenotypes, indicating that this fusion protein is biologically active. In these transgenic plants, *PR* gene expression was observed only in the presence of an SAR-inducer even though NPR1–GFP was expressed constitutively, suggesting that the fusion protein also requires activation to be functional as the endogenous NPR1 protein. Upon SAR induction, NPR1–GFP was observed exclusively in the nucleus (M. Kinkema, W. Fan & X. Dong, unpublished data). The activity of the NLS found in NPR1 was confirmed by site-directed mutagenesis followed by characterization in transgenic plants. Mutations in the NLS resulted in cytoplasmically localized npr1–GFP proteins which are inactive in regulating *PR* gene expression and SAR. Furthermore, we demonstrated that there is a direct correlation between the intensity of NPR1–GFP nuclear fluorescence and the induction of downstream *PR* genes (Fig. 2). At 0.1 mM SA, little nuclear NPR1–GFP fluorescence was observed, correlating with the lack of *PR1* gene expression; while at 0.3 mM SA, strong nuclear fluorescence was detected, concurring with a dramatic induction of the *PR1* gene. These data lead us to believe that NPR1 may regulate *PR* gene transcription directly. However, no obvious DNA binding domain has been identified in NPR1. It is possible that NPR1 acts as a coactivator to regulate *PR* gene promoters through interaction with a transcription factor.

    Consistent with this hypothesis, the TGA class of bZIP transcription factors was found to interact specifically with NPR1 in a yeast two-hybrid screen performed in my lab (Zhang et al 1999) and other labs. The specificity of the interaction between NPR1 and TGA transcription factors was demonstrated using deletion and point mutants of NPR1 as bait. A single amino acid change corresponding to *npr1-1* of

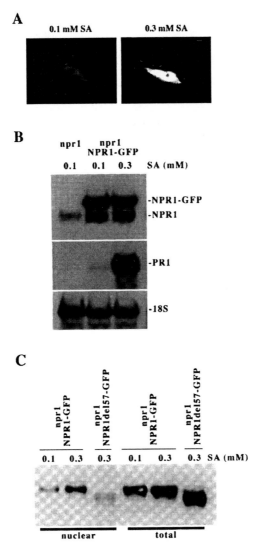

FIG. 2. Increased nuclear accumulation of NPR1–GFP is correlated with elevated levels of *PR* gene expression. (A) Representative images of nuclear NPR1–GFP fluorescence in leaf mesophyll cells of *Arabidopsis* seedlings grown on MS media containing either 0.1 mM or 0.3 mM SA. (B) Northern blot of RNA from *npr1* mutant or an *npr1* line expressing NPR1–GFP. Seedlings were grown on MS media containing either 0.1 mM or 0.3 mM SA. The blot was probed for *NPR1*, *PR1*, and *18S rRNA*. (C) Western blot of total protein and nuclear-fractionated protein. Protein was isolated from transgenic seedlings expressing either NPR1–GFP or NPR1del57–GFP. Seedlings were grown for 12 days on MS media containing either 0.1 mM or 0.3 mM SA. The blot was probed with antibodies to GFP. NPR1del57–GFP is a deletion mutant of NLS that is cytoplasmically localized and biologically inactive.

A. Y2H with WT and mutant NPR1 as bait   **B. Western blot of NPR1 in Y2H**

Leu drop-out   X-gal

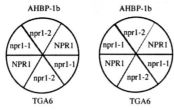

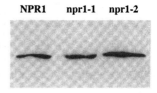

FIG. 3. Interactions between mutant npr1 protein and the bZIP transcription factors. (A) Yeast two-hybrid assay performed by using *npr1-1* and *npr1-2* point mutations as bait. Growth on a leu drop-out medium and development of a blue colour on X-gal plate are indicators of NPR1-bZIP interaction. (B) Western blot of the expression of *npr1* mutant proteins in yeast.

the ankyrin-repeat domain or *npr1-2* of the BTB/POZ domain completely abolished the ability of NPR1 to interact with the TGA transcription factors even though they had little effect on the accumulation of the mutant proteins in yeast (Fig. 3; Zhang et al 1999). Truncated NPR1 missing the C-terminal third of the protein but still maintaining the ankyrin-repeat and BTB/POZ domains showed the same interacting activity as the full-length NPR1 (Zhang et al 1999).

Our finding that NPR1 interacts specifically with the TGA subclass of bZIP transcription factors brings together the genetic data on the function of NPR1 in SAR with the results of previous molecular studies of SA-regulated genes. For example, the promoter of the *Arabidopsis PR1* gene has been thoroughly analysed using deletion and linker scanning mutagenesis as well as *in vivo* footprinting analysis (Lebel et al 1998). Through these analyses, two INA-responsive elements have been defined. One element at −610 is similar to the recognition sequence of NF-κB, while the other promoter element around residue −640 contains a CGTCA motif (the complementary sequence is TGACG) which is present in the *as1* element. The CGTCA motif was shown by linker-scanning mutagenesis to be essential for both SA- and INA-induced *PR1* gene expression. The *as1* element is a known binding site for TGA transcription factors. We used this SA- and INA-responsive *PR1* promoter element containing the *as1*-like sequence in a gel mobility shift assay. We found that purified TGA

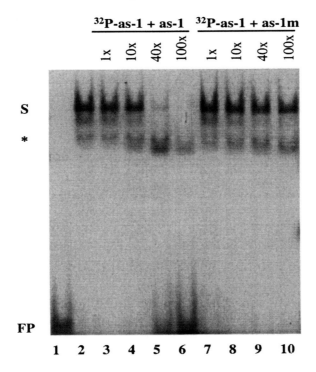

FIG. 4. AHBP-1b interacts specifically with the *Arabidopsis PR-1* promoter sequence containing an *as1*-like element. All binding reactions contain 4×10^4 cpm ^{32}P-labelled probe, incubated with 1 µg of either a control protein preparation (lane 1) or AHBP-1b (lanes 2–10), in the absence (lane 1 and 2) or presence (lanes 3–6) or non-specific (lanes 7–10) competitor probes. In lanes 3–6, 1×, 10×, 40× and 100× molar excess of the unlabelled probe containing point mutations in the *as1*-like element were added, respectively. FP, free probe; S, specific banding probe; *, non-specific banding.

transcription factor AHBP-1b can bind specifically to this promoter element *in vitro* (Fig. 4; Zhang et al 1999).

Beside the positive promoter elements, *as1* and the NF-κB binding site, a negative promoter element was also found in the *PR1* promoter (Lebel et al 1998). Mutations of this region by linker scanning resulted in an upward shift in both basal and induced gene expression. This implies that multiple transcription factors may bind to *PR* gene promoters and the induction of these genes may require turning on a transcription activator(s) and turning off a transcription repressor(s).

The existence of a negative regulator for SAR-related gene expression was also demonstrated by our genetic studies of the *sni1* mutant (suppressor *npr1*, inducible) (Li et al 1999). In this recessive mutant, SA- or INA-induced *PR* gene expression and resistance is restored in the *npr1* background. Cloning and sequencing analysis showed that the *sni1* mutation causes a frameshift in the N-terminal half of the protein and therefore a possible knockout of the protein function. Transformation of wild-type SNI1 into *npr1sni1* suppressed the induced gene expression and resistance of the *sni1* mutant, reverting the phenotype of the resulting transgenic plants back to that of *npr1*. These genetic data indicate that wild-type SNI1 is a negative regulator of SAR and the function of NPR1 is required for its inactivation. In the *npr1* mutant, SNI1 cannot be inactivated therefore no SAR induction occurs. The *sni1* mutation impairs the protein function and eliminates the requirement for NPR1. At present, it is not known whether NPR1 interacts physically with SNI1 because the result from a yeast two-hybrid assay was negative (Y. Zhang & X. Dong, unpublished data). It is worth noting that in *npr1sni1*, SAR is inducible rather than constitutive suggesting that an NPR1-independent activation event is also required for SAR-induction in parallel with the NPR1–SNI1 regulation (Li et al 1999). Intriguingly, SNI1 encodes a novel protein with some homology to mouse RB (retinoblastoma), a tumour suppressor that represses the expression of cell cycle genes by either inhibiting transcription activators such as E2F or recruitment of histone deacetylase.

Taken together, these data indicate that NPR1 is a key positive regulator of SAR. The activity of NPR1 is regulated post-transcriptionally. Our recent data suggest that NPR1 functions in the nucleus to regulate the *PR* genes. The activated NPR1 may exert its function by forming a protein complex with TGA transcription factors and/or by inactivating the transcription repressor, SNI1, to induce *PR* gene expression and SAR.

References

Bent A 1996 Plant disease resistance genes: function meets structure. Plant Cell 8:1757–1771

Cao H, Bowling, SA, Gordon AS, Dong X 1994 Characterization of an *Arabidopsis* mutant that is nonresponsive to inducers of systemic acquired resistance. Plant Cell 6:1583–1592

Cao H, Glazebrook J, Clarke JD, Volko S, Dong X 1997 The Arabidopsis *NPR1* gene that controls systemic acquired resistance encodes a novel protein containing ankyrin repeats. Cell 88:57–63

Cao H, Li X, Dong X 1998 Generation of broad-spectrum disease resistance by overexpression of an essential regulatory gene in systemic acquired resistance. Proc Natl Acad Sci USA 95:6531–6536

Delaney TP, Friedrich L, Ryals JA 1995 *Arabidopsis* signal transduction mutant defective in chemically and biologically induced disease resistance. Proc Natl Acad Sci USA 92:6602–6606

Enyedi AJ, Yalpani N, Silverman P, Raskin I 1992 Localization, conjugation and function of salicylic acid in tobacco during the hypersensitive reaction to tobacco mosaic virus. Proc Natl Acad Sci USA 89:2480–2484

Flor HH 1971 Current status of gene-for-gene concept. Annu Rev Phytopathol 9:275–296

Gaffney T, Friedrich L, Vernooij B et al 1993 Requirement of salicylic acid for the induction of systemic acquired resistance. Science 261:754–756

Glazebrook J, Rogers EE, Ausubel FM 1996 Isolation of *Arabidopsis* mutants with enhanced disease susceptibility by direct screening. Genetics 143:973–982

Hammond-Kosack K, Jones JDG 1996 Resistance gene-dependent plant defense responses. Plant Cell 8:1773–1791

Hammond-Kosack KE, Jones JDG 1997 Plant disease resistance genes. Annu Rev Plant Physiol Plant Mol Biol 48:575–607

Kuc J 1982 Induced immunity to plant disease. Bioscience 32:854–860

Lebel E, Heifetz P, Thorne L, Uknes S, Ryals J, Ward E 1998 Functional analysis of regulatory sequences controlling PR-1 gene expression in *Arabidopsis*. Plant J 16:223–234

Li X, Zhang Y, Clarke JD, Li Y, Dong X 1999 Identification and cloning of a negative regulator of systemic acquired resistance, SNI1, through a screen for suppressors of *npr1-1*. Cell 98:329–339

Métraux JP, Ahl-Goy P, Staub T, Speich J, Steinemann A, Ryals J, Ward E 1991 Induced resistance in cucumber in response to 2,6-dichloroisonicotinic acid and pathogens. In: Hennecke H, Verma DPS (eds) Advances in molecular genetics of plant–microbe interactions, vol 1. Kluwer Academic Publishers, Dordrecht, The Netherlands, p 432–439

Meuwly P, Buchala A, Mölders W, Métraux JP 1995 Local and systemic biosynthesis of salicylic acid in infected cucumber plants. Plant Physiol 109:1107–1114

Ross AF 1961 Systemic acquired resistance induced by localized virus infections in plants. Virology 14:340–358

Ryals JA, Neuenschwander UH, Willits MG, Molina A, Steiner H-Y, Hunt MD 1996 Systemic acquired resistance. Plant Cell 8:1809–1819

Ryals JA, Weymann K, Lawton K et al 1997 The *Arabidopsis* NIM1 protein shows homology to the mammalian transcription factor inhibitor IκB. Plant Cell 9:425–439

Shah J, Tsui F, Klessig DF 1997 Characterization of a salicylic acid-insensitive mutant (*sai1*) of *Arabidopsis thaliana*, identified in a selective screen utilizing the SA-inducible expression of the *tms2* gene. Mol Plant Microbe Interact 10:69–78

Staskawicz BJ, Ausubel FM, Baker BJ, Ellis JG, Jones JDG 1995 Molecular genetics of plant disease resistance. Science 268:661–667

Uknes S, Mauch-Mani B, Moyer M et al 1992 Acquired resistance in *Arabidopsis*. Plant Cell 4:645–656

White RF 1979 Acetylsalicylic acid (aspirin) induces resistance to tobacco mosaic virus in tobacco. Virology 99:410–412

Zhang Y, Fan W, Kinkema M, Li X, Dong X 1999 Interaction of NPR1 with basic leucine zipper protein transcription factors that bind sequences required for salicylic acid induction of *PR-1* gene. Proc Natl Acad Sci USA 96:6523–6528

DISCUSSION

Salmeron: My question relates to your hypothesis that the disease resistance in the NPR1 over-expressors is due to the combined moderate induction levels of all the PR genes. As you said, you didn't see a strong induction. What is the minimum level of SA or INA to achieve that level of induction, and is that enough chemical to provide disease resistance?

Dong: We don't know. We have different double and triple mutants in which we have measured SA levels. I asked my student to see whether there is a threshold level of SA that is required for resistance, but I don't think that we know this.

Leung: What is the phenotype of the single *sni1* mutation?

Dong: They are smaller than wild-type. When this negative regulator is mutated, the system is leaky and the growth is slightly stunted.

Leung: Am I right in thinking that if there is just a single *sni1* mutation, the plant requires less SA for resistance?

Dong: Yes, 10–100-fold less SA is needed.

Leung: Have you any suggestions for how we could screen for NPR equivalents in rice?

Dong: In rice, you can induce resistance with BTH (benzo[1,2,3]thiadiazole-7-carbothioic acid S-methyl ester), so you could do the same experiment that Terry Delaney did at Novartis, spraying BTH on the mutagenized population looking for loss of BTH-induced resistance (Delaney et al 1995).

Bennett: In your model, you have NPR1 acting in the nucleus. However, many ankyrin repeat proteins are actually cytosolic. Is there a possibility that NPR is bound to the plasma membrane and that the nuclear localization signal is cryptic until it is activated by the pathogen? By putting GUS on the C-terminus you may have prevented the nuclear localization signal from becoming cryptic, so it goes straight into the nucleus.

Dong: Our nuclear localization study is incomplete because we don't know where the protein is before induction: we don't see the NPR1–GFP before induction. Western blotting using anti-GFP antibody showed that the same amount of fusion protein is made before and after induction, but a much stronger band is visible after induction in the nuclear fraction. It certainly moves from somewhere to the nucleus, but we don't know from where exactly. Our speculation is that it is probably from the cytoplasm.

G.-L. Wang: After you cloned the gene, your lab and others used the yeast two-hybrid system to demonstrate that TGA6 is important in activation of the PR1 protein. When you do the screening you use GUS and the PR1 promoter. Why haven't you identified any mutants in the *TGA* gene?

Dong: One explanation is that this is probably because it is a gene family, so when you mutate one *TGA* there are still other functional *TGA*s present. If this is the case, straightforward genetic screens may not lead to the identification of all the genes involved in SAR regulation.

Parker: You and others have shown that NPR1 functions in an SA-independent pathway that may be related to jasmonic acid-mediated signalling, or other signals. Do you think that your results would be consistent with the idea that NPR1 might be associating with multiple transcription factors, and its association may depend on the particular input signal? Is that reflected in the yeast two-hybrid analysis that you and others have carried out?

Dong: Yes, that is probably the case. Expression studies of TGA transcription factor genes have shown that one or two of them have expression patterns almost identical to that of NPR1. NPR1 may regulate a very different set of genes involved in SAR or induced systemic resistance by forming complexes with different partners. This is something that I am very interested in.

Reference

Delaney T, Friedrich L, Ryals J 1995 *Arabidopsis* signal transduction mutant defective in chemically and biologically induced disease resistance. Proc Natl Acad Sci USA 92:6602–6606

Improving plant drought, salt and freezing tolerance by gene transfer of a single stress-inducible transcription factor

Kazuko Yamaguchi-Shinozaki and Kazuo Shinozaki*

*Biological Resources Division, Japan International Research Center for Agricultural Sciences (JIRCAS), Ministry of Agriculture, Forestry and Fisheries, 2-1 Ohwashi, Tsukuba, Ibaraki 305-8686, and *Laboratory of Plant Molecular Biology, The Institute of Physical and Chemical Research (RIKEN), 3-1-1 Koyadai, Tsukuba, Ibaraki 305-0074, Japan*

Abstract. Plant productivity is greatly affected by environmental stresses such as drought, salt loading and freezing. We reported that a *cis*-acting promoter element, the dehydration response element (DRE), plays an important role in regulating gene expression in response to these stresses in *Arabidopsis*. The transcription factor DREB1A specifically interacts with the DRE and induces expression of stress tolerance genes. We show here that overexpression of the cDNA encoding DREB1A in transgenic *Arabidopsis* plants activated the expression of many of theses stress tolerance genes under normal growing conditions and resulted in improved tolerance to drought, salt loading and freezing. However, use of the strong constitutive 35S cauliflower mosaic virus (CaMV) promoter to drive expression of DREB1A also resulted in severe growth retardation under normal growing conditions. In contrast, expression of DREB1A from the stress-inducible rd29A promoter gave rise to minimal effects on plant growth while providing an even greater tolerance to stress conditions than did expression of the gene from the CaMV promoter. As the DRE-related regulatory element is not limited to *Arabidopsis* the DREB1A cDNA and the rd29A promoter may be useful for improving the stress tolerance of agriculturally important crops by gene transfer.

2001 Rice biotechnology: improving yield, stress tolerance and grain quality. Wiley, Chichester (Novartis Foundation Symposium 236) p 176–189

Drought, salt loading and freezing are environmental conditions that cause adverse effects on the growth of plants and the productivity of crops. Plants respond to these stresses at molecular and cellular levels as well as the physiological level. Expression of a variety of genes is induced by these stresses (Shinozaki & Yamaguchi-Shinozaki 1996, Thomashow 1994). The products of these genes are thought to function not only in stress tolerance but also in the regulation of gene expression and signal transduction in stress response (Shinozaki & Yamaguchi-Shinozaki 1997).

Genetic engineering may be useful for improving plant stress tolerance, and several different approaches have recently been attempted (Holmberg & Bulow 1998). The genes selected for transformation were those involved in encoding enzymes required for the biosynthesis of various osmoprotectants (Tarczynski & Bohnert 1993, Kavi Kishor et al 1995, Hayashi et al 1997). Other genes that have been selected for transformation include those that encoded enzymes for modifying membrane lipids, LEA protein and detoxification enzyme (Kodama et al 1994, Ishizaki-Nishizawa et al 1996, Xu et al 1996, McKersie et al 1996). In all these experiments, a single gene for a protective protein or an enzyme was overexpressed under the control of the 35S cauliflower mosaic virus (CaMV) constitutive promoter in transgenic plants, although a number of genes have been shown to function in environmental stress tolerance and response. The genes encoding protein factors that are involved in regulation of gene expression, signal transduction and function in the stress response are promising candidates for the genetic engineering of stress tolerance as they can regulate a range of stress-inducible genes.

Drought is one of the most severe environmental stresses and affects almost all plant functions. Abscisic acid (ABA) is produced under water deficit conditions and plays important roles in tolerance against drought. Most of the drought-inducible genes that have been studied to date are also induced by ABA (Shinozaki & Yamaguchi-Shinozaki 1997). Several reports have described genes that are induced by dehydration but are not responsive to exogenous ABA treatments (Shinozaki & Yamaguchi-Shinozaki 1997). These findings suggest the existence of ABA-independent as well as ABA-dependent signal transduction cascades between the initial signal of drought stress and the expression of specific genes (Shinozaki & Yamaguchi-Shinozaki 1997). To understand the molecular mechanisms of gene expression in response to drought stress, we have analysed in detail cis- and trans-acting elements that function in ABA-independent and ABA-responsive gene expression by drought stress (Shinozaki & Yamaguchi-Shinozaki 1997). In this paper we summarize the recent progress of our research on cis- and trans-acting factors involved in ABA-independent gene expression in drought stress response. We also report stress tolerance of transgenic plants that overexpress a single gene for a stress-inducible transcription factor using Arabidopsis as a model.

The function of water stress-inducible genes

A variety of genes are induced by drought-stress, and the functions of their gene products have been predicted from sequence homology with known proteins.

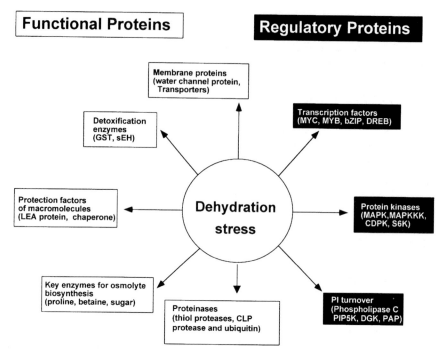

FIG. 1. Drought-stress inducible genes and their possible functions in stress tolerance and response. Gene products are classified into two groups. The first group includes proteins that probably function in stress tolerance (functional proteins); the second group contains protein factors involved in further regulation of signal transduction and gene expression that probably function in stress response (regulatory proteins).

Genes induced during drought stress conditions are thought to function not only in protecting cells from dehydration by the production of important metabolic proteins, but also in the regulation of genes for signal transduction in the drought stress response (Shinozaki & Yamaguchi-Shinozaki 1996, 1997). Thus, these gene products are classified into two groups (Fig. 1).

    The first group includes proteins that probably function in stress tolerance (Shinozaki & Yamaguchi-Shinozaki 1997): water channel proteins involved in the movement of water through membranes, the enzymes required for the biosynthesis of various osmoprotectants (sugars, proline and betaine), proteins that may protect macromolecules and membranes (LEA protein, osmotin, antifreeze protein, chaperones and mRNA binding proteins), proteases for protein turnover (thiol proteases, Clp protease, and ubiquitin) and the detoxification enzymes (glutathione S-transferase, soluble epoxide hydrolase, catalase, superoxide dismutase and ascorbate peroxidase). Some of the stress-inducible genes that encode proteins such as a key enzyme for proline biosynthesis, have been overexpressed in transgenic plants to produce a stress-tolerant phenotype of the plants (Kavi Kishor et al 1995).

The second group contains protein factors involved in the further regulation of signal transduction and gene expression that probably function in the stress response: protein kinases, transcription factors and enzymes involved in phospholipid metabolism (Shinozaki & Yamaguchi-Shinozaki 1997). It is important for us to elucidate the role of these regulatory proteins for understanding plant responses to water stress and for improving the tolerance of plants by gene transfer. The existence of a variety of drought-inducible genes suggests complex responses of plants to drought stress. Their gene products are involved in drought stress tolerance and stress responses.

Expression of dehydration-induced genes in response to environmental stresses and ABA

We analysed the expression patterns of genes induced by drought by RNA gel-blot analysis. The results indicated broad variations in the timing of induction of these genes under drought conditions. Most of the drought-inducible genes respond to treatment with exogenous ABA, but others do not (Shinozaki & Yamaguchi-Shinozaki 1996, 1997). Therefore, there are not only ABA-dependent but also ABA-independent regulatory systems of gene expression under drought stress. Analysis of the expression of ABA-inducible genes revealed that several genes require protein biosynthesis for their induction by ABA, suggesting that at least two independent pathways exist between the production of endogenous ABA and gene expression during stress.

As shown in Fig. 2, we identified at least four independent signal pathways which function under drought conditions: two are ABA-dependent (pathways I and II) and two are ABA-independent (pathways III and IV). One of the ABA-dependent pathways overlaps with that of the cold response (pathway IV). One of the ABA-independent pathways requires protein biosynthesis (pathway II) (Shinozaki & Yamaguchi-Shinozaki 1996, 1997). The existence of complex signal transduction pathways in the drought response gives a molecular basis for the complex physiological responses of plants to drought stress.

Identification of a *cis*-acting element, DRE, involved in drought-responsive gene expression

A number of genes are induced by drought, salt and cold in *aba* (ABA-deficient) or *abi* (ABA-insensitive) *Arabidopsis* mutants. This suggests that these genes do not require ABA for their expression under cold or drought condition. Among these

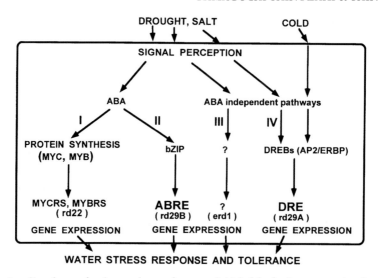

FIG. 2.   Signal transduction pathways between initial dehydration stress signal and gene expression. There are at least four signal transduction pathways: two are ABA-dependent (I and II) and two are ABA-independent (III and IV). Protein synthesis is necessary for one of the ABA-dependent signal pathways (I). ABRE is involved in one of the ABA-dependent pathways (II). In one of the ABA-independent pathways, DRE is involved in the regulation of genes not only by drought and salt but also by cold stress (IV). Another ABA-independent pathway is controlled by drought and salt, but not by cold (III).

genes, the expression of a drought-inducible gene for rd29A/lti78/cor78 was extensively analysed (Yamaguchi-Shinozaki & Shinozaki 1994). At least two separate regulatory systems function in gene expression during drought and cold stress; one is ABA-independent (Fig. 2, pathway IV) and the other is ABA-dependent (pathway II).

To analyse the *cis*-acting elements involved in the ABA-independent gene expression of rd29A, we constructed chimeric genes with the rd29A promoter fused to the β-glucuronidase (*GUS*) reporter gene and transformed *Arabidopsis* and tobacco plants with these constructs. The *GUS* reporter gene driven by the rd29A promoter was induced at significant levels in transgenic plants by dehydration, low-temperature, high-salt or ABA treatment (Yamaguchi-Shinozaki & Shinozaki 1993). The deletion, the gain-of-function and the base substitution analysis of the promoter region of *rd29A* gene revealed that a 9 bp conserved sequence, TACCGACAT (DRE, Dehydration Responsive Element), is essential for the regulation of the expression of *rd29A* under drought conditions. Moreover, DRE has been demonstrated to function as a *cis*-acting element involved in the induction of *rd29A* by either low-temperature or high-salt stress (Yamaguchi-Shinozaki & Shinozaki 1994). Therefore, DRE seems to be a *cis*-acting element involved in gene induction by dehydration, high-salt, or low temperature, but does not function as an ABA-responsive element in the induction of *rd29A*.

Important roles of the DRE binding proteins during drought and cold stresses

Two cDNA clones that encode DRE binding proteins, DREB1A and DREB2A, were isolated by using the yeast one-hybrid screening technique (Liu et al 1998). The deduced amino acid sequences of DREB1A and DREB2A showed significant sequence similarity with the conserved DNA binding domains found in the EREBP and APETALA2 proteins that function in ethylene-responsive expression and floral morphogenesis, respectively (Okamuro et al 1997, Ohme-Takagi & Shinshi 1995). Each DREB protein contained a basic region in its N-terminal region that might function as a nuclear localization signal and an acidic C-terminal region that might act as an activation domain for transcription. These data suggest that each DREB cDNA encodes a DNA binding protein that might function as a transcriptional activator in plants.

We examined the ability of the DREB1A and DREB2A proteins expressed in *Escherichia coli* to bind the wild-type or mutated DRE sequences using the gel retardation method (Liu et al 1998). The results indicate that the binding of these two proteins to the DRE sequence is highly specific. To determine whether the DREB1A and DREB2A proteins are capable of transactivating DRE-dependent transcription in plant cells, we performed transactivation experiments using protoplasts prepared from *Arabidopsis* leaves. Coexpression of the DREB1A or DREB2A proteins in protoplasts transactivated the expression of the *GUS* reporter gene. These results suggest that DREB1A and DREB2A proteins function as transcription activators involved in the cold- and dehydration-responsive expression, respectively, of the *rd29A* gene (Fig. 3) (Liu et al 1998).

We isolated cDNA clones encoding two DREB1A homologues (named DREB1B and DREB1C). The DREB1B clone was identical to CBF1 (Stockinger et al 1997). We also isolated cDNA clones encoding a DREB2A homologue and named it DREB2B. Expression of the *DREB1A* gene and its two homologues was induced by low-temperature stress, whereas expression of the *DREB2A* gene and its single homologue was induced by dehydration (Liu et al 1998, Shinwari et al 1998). These results indicate that two independent families of DREB proteins, DREB1 and DREB2, function as *trans*-acting factors in two separate signal transduction pathways under low-temperature and dehydration conditions, respectively (Fig. 3) (Liu et al 1998).

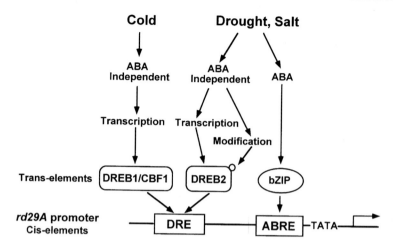

FIG. 3. A model of the induction of the *rd29A* gene and *cis-* and *trans*-acting elements involved in stress-responsive gene expression. Two *cis*-acting elements, DRE and ABRE, are involved in the ABA-independent and ABA-responsive induction of *rd29A*, respectively. Two types of different DRE binding proteins, DREB1 and DREB2, separate two different signal transduction pathways in response to cold and drought stresses, respectively. DREB1s/ CBF1 are transcriptionally regulated whereas DREB2s are controlled post-translationally as well as transcriptionally. ABRE binding proteins encode bZIP transcription factors.

Analysis of the in roles of DREB1a and DREB2a by using transgenic plants

We generated transgenic plants in which *DREB1A* or *DREB2A* cDNAs were introduced to overproduce DREB proteins to analyse the effects of overproduction of DREB1A and DREB2A proteins on the expression of the target *rd29A* gene. *Arabidopsis* plants were transformed with vectors carrying fusions of the enhanced CaMV 35S promoter and the *DREB1A* (*35S:DREB1A*) or *DREB2A* (*35S:DREB2A*) cDNAs in the sense orientation (Liu et al 1998, Mituhara et al 1996). All of the transgenic plants carrying the *35S:DREB1A* transgene (the *35S:DREB1A* plants) showed growth-retardation phenotypes under normal growth conditions. The *35S:DREB1A* plants showed variations in phenotypic changes in growth retardation that may have been due to the different levels of expression of the *DREB1A* transgenes for the position effect (Liu et al 1998).

To analyse whether overproduction of the DREB1A protein caused the expression of the target gene in unstressed plants, we compared the expression of the *rd29A* gene in control plants carrying the pBI121 vector. Transcription of the *rd29A* gene was low in the unstressed wild-type plants but high in the unstressed *35S:DREB1A* plants (Liu et al 1998). The level of the *rd29A* transcripts under the unstressed control condition was found to depend on the level of the DREB1A transcripts (Liu et al 1998). To analyse whether overproduction of the DREB1A protein caused the expression of other target genes, we evaluated the expression of its target stress-inducible genes. In the *35S:DREB1A* plants the *kin1*, *cor6.6*/*kin2*, *cor15a*, *cor47*/*rd17* and *erd10* genes were expressed strongly under unstressed control conditions, as was the *rd29A* gene (Shinozaki & Yamaguchi-Shinozaki 1997, Kiyosue et al 1994).

In contrast, the transgenic plants carrying the *35S:DREB2A* transgene (the *35S:DREB2A* plants) showed little phenotypic change. In *35S:DREB2A* transgenic plants, the *rd29A* mRNA did not accumulate significantly, although the *DREB2A* mRNA accumulated even under unstressed conditions (Liu et al 1998). Expression of the DREB2A protein is not sufficient for the induction of the target stress-inducible gene. Modification, such as phosphorylation of the DREB2A protein, seems to be necessary for its function in response to dehydration (Fig. 3). However, DREB1 proteins can function without modification.

Drought, salt and freezing stress tolerance in transgenic plants

The tolerance to freezing and dehydration of the transgenic plant was analysed using the *35S:DREB1A* plants grown in pots at 2 °C for 3 weeks. When plants were exposed to a temperature of −6 °C for 2 days, returned to 22 °C, and grown for 5 days, all of the wild-type plants died, whereas the *35S:DREB1A* plants survived at high frequency (Liu et al 1998, Kasuga et al 1999). Freezing tolerance was correlated with the level of expression of the stress-inducible genes under unstressed control conditions (Fig. 4; between 80 and 30%, survival) (Liu et al 1998, Kasuga et al 1999).

To test whether the introduction of the *DREB1A* gene enhances tolerance to dehydration stress, we did not water the plants for 2 weeks. Although all of the wild-type plants died within 2 weeks, between 70 and 20% of the *35:SDREB1A* plants survived and continued to grow after re-watering. Drought tolerance was also dependent on the level of expression of the target genes in the *35S:DREB1A* plants under unstressed conditions (Fig. 4) (Liu et al 1998, Kasuga et al 1999).

Overexpression of the *DREB1A* cDNA, driven by the constitutive 35S CaMV promoter in transgenic plants, activated strong expression of the target stress-inducible genes under unstressed conditions, which, in turn, increased tolerance of freezing, salt and drought stresses(Liu et al 1998, Kasuga et al 1999). Jaglo-Ottosen et al (1998) reported that CBF1 overexpression also enhances freezing tolerance. However, the overexpression of stress-inducible genes controlled by the DREB1A protein caused severe growth retardation under normal growth conditions (Liu et al 1998, Kasuga et al 1999).

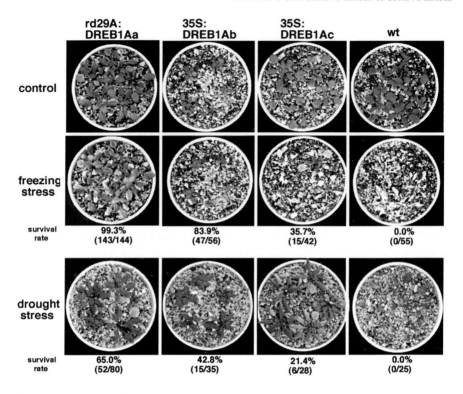

FIG. 4. Freezing and drought stress tolerance of the *35S:DREB1Ab*, *35S:DREB1Ac* and *rd29A:DREB1Aa* transgenic plants. The stress treatments were conducted as described in the text. Control, 3 week old plants growing under normal conditions; freezing, plants exposed to a temperature of −6 °C for 2 days and returned to 22 °C for 5 days; drought, water withheld for 2 weeks. Percentages of surviving plants and numbers of surviving plants per total tested plants are given under the pictures.

To resolve the problem of growth retardation, we used the stress-inducible rd29A promoter to cause overexpression of *DREB1A* in transgenic plants (*rd29A:DREB1A* plants) (Kasuga et al 1999). Because the rd29A promoter was stress-inducible and contained binding sites for the DREB1A protein, it did not cause expression of the *DREB1A* transgene at high levels under unstressed conditions; instead, it rapidly amplified expression of the *DREB1A* transgene only under dehydration, salt and low-temperature stress. The *rd29A:DREB1A* plants revealed strong stress tolerance even though their growth retardation under normal growing conditions was not significant. Moreover, the growth and the productivity of these plants were almost the same as those of the wild-type plants under normal growing conditions (Kasuga et al 1999).

On the contrary, the *rd29A:DREB1A* transgenic plants are more tolerant to the stresses than the *35S:DREB1A* plants that exhibited growth retardation under normal growing conditions (Fig. 4) (Kasuga et al 1999). As the *rd29A* gene is one of the target genes of the DREB1A protein, the rd29A promoter is more suitable for the tissue-specific expression of the *DREB1A* gene in plants than the 35S CaMV promoter. In the *rd29A:DREB1A* plants, the target gene products seem to be strongly accumulated in the same tissues that express the products under stress conditions. These results indicate that combination of the *DREB1A* cDNA with the rd29A promoter would be quite useful for improving drought, salt and freezing-stress tolerance in transgenic plants.

In a previous studies, we showed that DRE also functions in gene expression in response to stress in tobacco plants (Yamaguchi-Shinozaki & Shinozaki 1993, 1994), which suggests the existence of similar regulatory systems in tobacco and other crop plants. DRE-related motifs have been reported in the promoter region of cold-inducible *Brassica napus* and wheat genes (Jiang et al 1996, Ouellet et al 1998). These observations suggest that both the *DREB1A* cDNA and the rd29A promoter can be used to improve the dehydration, salt and freezing tolerance of crops by gene transfer.

References

Hayashi H, Mustardy L, Deshnium P, Ida M, Murata N 1997 Transformation of *Arabidopsis thaliana* with the codA gene for choline oxidase; accumulation of glycinebetaine and enhanced tolerance to salt and cold stress. Plant J 12:133–142

Holmberg N, Bulow L 1998 Improving stress tolerance in plants by gene transfer. Trends Plant Sci 3:61–66

Ishizaki-Nishizawa O, Fujii T, Azuma M et al 1996 Low-temperature resistance of higher plants is significantly enhanced by a nonspecific cyanobacterial desaturase. Nat Biotechnol 14:1003–1006

Jaglo-Ottosen KR, Gilmour SJ, Zarka DG, Schabenberger O, Thomashow MF 1998 *Arabidopsis* CBF1 overexpression induces *cor* genes and enhances freezing tolerance. Science 280:104–106

Jiang C, Iu B, Singh J 1996 Requirement of a CCGAC *cis*-acting element for cold induction of the *BN115* gene from winter *Brassica napus*. Plant Mol Biol 30:679–684

Kasuga M, Liu Q, Miura S, Yamaguchi-Shinozaki K, Shinozaki K 1999 Improving plant drought, salt, and freezing tolerance by gene transfer of a single stress-inducible transcription factor. Nat Biotechnol 17:287–291

Kavi Kishor PB, Hong Z, Miao G-U et al 1995 Overexpression of D^1-pyrroline-5-carboxylate synthetase increases proline production and confers osmotolerance in transgenic plants. Plant Physiol 108:1387–1394

Kiyosue T, Yamaguchi-Shinozaki K, Shinozaki K 1994 Characterization of two cDNAs (ERD10 and ERD14) corresponding to genes that respond rapidly to dehydration stress in *Arabidopsis thaliana*. Plant Cell Physiol 35:225–231

Kodama H, Hamada T, Horiguchi G et al 1994 Genetic enhancement of cold tolerance by expression of a gene for chloroplast ω-3 fatty acid desaturase in transgenic tobacco. Plant Physiol 105:601–605

Liu Q, Kasuga M, Sakuma Y et al 1998 Two transcription factors, DREB1 and DREB2, with an EREBP/AP2 DNA binding domain, separate two cellular signal transduction pathways in drought- and low temperature-responsive gene expression, respectively, in *Arabidopsis*. Plant Cell 10:1391–1406

McKersie BD, Bowley SR, Harjanto E et al 1996 Water-deficit tolerance and field performance of transgenic alfalfa overexpressing superoxide dismutase. Plant Physiol 111:1177–1181

Mituhara I, Ugaki M, Hirochika H et al 1996 Efficient promoter cassettes for enhanced expression of foreign genes in dicotyledonous and monocotyledonous plants. Plant Cell Physiol 37:49–59

Ohme-Takagi M, Shinshi H 1995 Ethylene-inducible DNA binding proteins that interact with an ethylene-responsive element. Plant Cell 7:173–182

Okamuro JK, Caster B, Villarroel R, Van Montagu M, Jofuku KD 1997 The AP2 domain of *APETALA2* defines a large new family of DNA binding proteins in *Arabidopsis*. Proc Natl Acad Sci USA 94:7076–7081

Ouellet F, Vazquez-Tello A, Sarhan F 1998 The wheat *wcs120* promoter is cold-inducible in both monocotyledonous and dicotyledonous species. FEBS Lett 423:324–328

Shinozaki K, Yamaguchi-Shinozaki K 1996 Molecular responses to drought and cold stress. Curr Opin Biotechnol 7:161–167

Shinozaki K, Yamaguchi-Shinozaki K 1997 Gene expression and signal transduction in water-stress response. Plant Physiol 115:327–334

Shinwari ZK, Nakashima K, Miura S et al 1998 An *Arabidopsis* gene family encoding DRE/CRT binding proteins involved in low-temperature-responsive gene expression. Biochem Biophys Res Commun 250:161–170

Stockinger EJ, Gilmour SJ, Thomashow MF 1997 *Arabidopsis thaliana* CBF1 encodes an AP2 domain-containing transcriptional activator that binds to the C-repeat/DRE, a *cis*-acting DNA regulatory element that stimulates transcription in response to low temperature and water deficit. Proc Natl Acad Sci USA 94:1035–1040

Tarczynski M, Bohnert H 1993 Stress protection of transgenic tobacco by production of the osmolyte mannitol. Science 259:508–510

Thomashow MF 1994 *Arabidopsis thaliana* as a model for studying mechanisms of plant cold tolerance. In: Meyrowitz E, Somerville C (eds) *Arabidopsis*. Cold Spring Harbor Laboratory Press, Cold Spring Harbor, NY, p 807–834

Xu D, Duan X, Wang B et al 1996 Expression of a late embryogenesis abundant protein gene, HVA1, from barley confers tolerance to water deficit and salt stress in transgenic rice. Plant Physiol 110:249–257

Yamaguchi-Shinozaki K, Shinozaki K 1993 Characterization of the expression of a desiccation-responsive *rd29* gene of *Arabidopsis thaliana* and analysis of its promoter in transgenic plants. Mol Gen Genet 236:331–340

Yamaguchi-Shinozaki K, Shinozaki K 1994 A novel *cis*-acting element in an *Arabidopsis* gene is involved in responsiveness to drought, low-temperature, or high-salt stress. Plant Cell 6: 251–264

DISCUSSION

Leach: In addition to the non-abiotic stresses, did you test any biological stresses on these plants? Were they protected?

Yamaguchi-Shinozaki: We haven't tried them yet. However, we found that one of the targets for DREB1A is a gene for heat shock protein, so now we are analysing the heat shock tolerance of these *DREB1A* transgenic plants.

Ku: Following on from this, I think it might be worth looking at how these transgenics respond to other stresses, such as pathogen stress. The recent work by Griffith's group at the University of West Ontario (Antikainen & Griffith 1997), found large amounts of cold-induced antifreeze protein in cold-tolerant cereal crops. The twist in the story is that these antifreeze proteins turn out to be pathogenesis-related (PR) proteins that have been recruited by the cereal crops for antifreeze functions. Perhaps your transgenics will also have enhanced pathogen resistance as well as cold and drought tolerance.

Yamaguchi-Shinozaki: We should try this.

Ku: Are you considering a strategy involving the expression of both transcriptional factors in the same plant? If so, would you expect additive effects on stress tolerance?

Yamaguchi-Shinozaki: Perhaps. We have also been working on the gene for proline synthetase. We got good results with the transgenic plants accumulating proline at higher levels. We are now trying to introduce genes for both DREB1A and proline synthetase into plants.

Okita: What do you think is the overall mechanism behind cold and drought tolerance? I am struck by the fact that constitutive expression of the genes involved greatly retards growth in these plants.

Yamaguchi-Shinozaki: Although some of the genes we have targeted affect plant growth, others don't. We have overexpressed rd29A without seeing any growth retardation.

Okita: When you overexpress *DREB2A*, is the plant just maintaining homeostasis of the cells, preventing growth from occurring and thereby allowing the plant to survive the stressful conditions? Or are you building some kind of mechanism into the plant that still allows it to grow under these stress conditions?

Yamaguchi-Shinozaki: We haven't tried this kind of experiment. We are now using the rd29A promoter, which is reversible. After the stress conditions, if the conditions are good, we can get enough seed from this plant.

Dong: Have you tried treating the rd29A-driven *DREB1A* transgenic with drought or cold at different developmental stages?

Yamaguchi-Shinozaki: We treat briefly with stress conditions that are severe enough that the wild-type plant would not survive, so we don't have a comparison.

Bennett: Am I to understand from your paper that the promoter of *DREB1A* is comparatively weak, and that when it is replaced by a stronger promoter with similar induction properties, there is a much higher level of DREB1 protein?

Yamaguchi-Shinozaki: Actually, we use the rd29A promoter to overexpress *DREB1A* in the transgenic plant because this promoter has a binding site for the DREB1 protein. This promoter can self-amplify. This promoter is the strongest one for expressing the DREB1A protein in *Arabidopsis*.

Salmeron: Have you looked for mutations in these transcription factors?

Yamaguchi-Shinozaki: DREB1A has two homologues. When DREB1A is disrupted, there is no difference in phenotype because the homologues compensate for its loss. If we wanted to make a mutant with a phenotype we would have to disrupt DREB1B and DREB1C also.

Mazur: How did you initially identify those stress- and drought-responsive genes?

Yamaguchi-Shinozaki: We used a differential screening method with cDNA probes prepared from dehydrated and non-dehydrated *Arabidopsis*.

Mazur: Did you pre-select the probes at all?

Yamaguchi-Shinozaki: We prepared a library using dehydration-treated *Arabidopsis*. It contained drought-inducible genes, and as a probe we used cDNA prepared from normal or dehydrated plants.

Li: Was the rd29A promoter originally isolated from a resistant genotype, or is it randomly isolated from the induced condition in a susceptible line?

Yamaguchi-Shinozaki: Originally we isolated the cDNAs for drought inducible genes. One of these was *rd29*. We isolated the promoter region of this gene using a genomic library.

Li: Was the cDNA library prepared from a drought-resistant or tolerant genotype?

Yamaguchi-Shinozaki: We used the colombia strain of *Arabidopsis*, which we subjected to dehydration.

Li: Does this line show any drought or salinity tolerance, or any co-tolerance to the stress?

Yamaguchi-Shinozaki: *Arabidopsis* shows some degree of drought tolerance.

Li: The reason I ask is that I want to know whether the original material shows a stronger tolerance as compared to the transgenic lines: is this strong promoter really from very good germplasm? If this is the case, we might want to screen with this method to find good promoters in rice or other species.

Yamaguchi-Shinozaki: We are also using the cowpea crop grown in a semi-arid region. This shows good drought stress tolerance. In this case we also isolated a similar cDNA.

Li: So what you suggest is that even from a germplasm without a very strong tolerance, we can identify good promoters or good alleles for these tolerant genes or quantitative trait loci (QTLs). This is an important issue. We need to know whether we need to look at the very best germplasm we have, or whether we can just randomly select from diverse sources of germplasm.

Dong: But if you already have such good germplasm you don't need molecular engineering.

Li: That is not the case: drought tolerance in rice is very complex. We will not be able to identify a single gene line in this case. We are very surprised that a single gene promoter can create such a high level of drought or salinity tolerance.

G.-L. Wang: You have shown that the rd29A promoter is much more effective than the 35S promoter. The next step will probably involve field-testing of these plants. Is this promoter also sensitive to temperature, sunlight and other environmental variables? In the greenhouse these can be controlled but it is hard to control them in the fields.

Yamaguchi-Shinozaki: We want to apply this technology to another crop plant. In this case we should try field testing. With *Arabidopsis* the *rd29A* gene is a strong promoter — perhaps even the strongest promoter in *Arabidopsis* under stress conditions. But when we tried this promoter in rice, we got different results. It is also inducible in the rice root, but we couldn't detect induction of the induced genes in the leaves. If we want to apply this technology to another crop plant, we have to isolate the original promoter for that type of crop plant. It may be possible to use DREB1A in rice also, but we need another promoter because of the different tissue specificity.

Reference

Antikainen M, Griffith M 1997 Antifreeze protein accumulation in freezing-tolerant cereals. Physiol Plant 99:423–432

Dissection of defence response pathways in rice

Jan E. Leach, Hei Leung* and Guo-Liang Wang†

Department of Plant Pathology, 4024 Throckmorton Plant Sciences Center, Kansas State University, Manhattan, KS 66506-5502, USA, *Division of Entomology and Plant Pathology, International Rice Research Institute, DAPO Box 7777, Metro Manila, Philippines, and †Department of Plant Pathology, 201 Kottman Hall, 2021 Coffey Rd, The Ohio State University, Columbus, OH 43210, USA

Abstract. The cloning of major resistance genes has led to a better understanding of the molecular biology of the steps for induction of resistance, yet much remains to be discovered about the downstream genes that collectively confer resistance, i.e. the defence response (DR) genes. We are dissecting the pathways contributing to resistance in rice by identifying a collection of mutants with deletions or other structural rearrangements in DR genes. The collection of rice mutants has been screened for many characters, including increased susceptibility or resistance to *Magnaporthe grisea* and *Xanthomonas oryzae* pv. *oryzae*. A collection of enhanced sequence tags (ESTs) and putative DR genes has been established to facilitate detection of mutants with deletions in DR genes. Arrays of DR genes will be used to create gene expression profiles of interesting mutants. Successful application of the mutant screen will have broad utility in identifying candidate genes involved in disease response and other metabolic pathways.

2001 Rice biotechnology: improving yield, stress tolerance and grain quality. Wiley, Chichester (Novartis Foundation Symposium 236) p 190–204

As we strive towards less dependence on chemicals for crop production, the need for durable genetic resistance against a broad spectrum of diseases becomes more compelling. This is particularly true in rice, which is the staple food for about half of the world's population. However, because of the variability of pathogens, a long-standing challenge in disease control in rice has been the identification and understanding of the basis of durable resistance. Through genetic analysis, we have progressed from a phenotypic description of durable resistance to its genetic interpretation. For example, durable resistance against rice blast is attributed to a combination of race-specific, qualitative resistance and non race-specific, quantitative resistance (Wang et al 1994). While the successful cloning of many major resistance genes has led to a better understanding of the molecular biology of qualitative resistance (Ellis & Jones 1998), much remains to be discovered about the genes that collectively confer quantitative resistance. If we are to manipulate these traits to sustain crop productivity, we need to know the function and role of particular genes conferring quantitative resistance. Rice, with its global significance as a food crop and as a major cereal for genomics and genetic research, provides an ideal model to investigate this fundamental question in crop protection.

One means to understand the contribution of a particular gene in resistance is to inactivate that gene and study the effects of the mutation on a phenotype. Strategies for mutagenesis have long been available; however, the ability to detect mutations with subtle phenotypes is limited. Due to technical advances in genomics research and the increasing availability of plant gene or expressed sequence tag (EST) sequences, approaches to overcome these limitations are realistic. In this paper, we describe progress on the development of a collection of rice deletion mutants and approaches to identify mutants with deletions in candidate resistance or defence response genes.

Qualitative and quantitative resistance

In understanding the molecular basis for durable resistance, two broad classes of genes are of interest; those involved in the recognition process (R or major genes) and those involved in the defence response (DR or minor genes). R genes code for recognition factors, and their recognition of pathogens stimulates a signal transduction pathway that activates the transcription of DR genes, which code for enzymes or structural components that impede pathogen ingress. Most R genes that have been characterized share common sequence motifs, such as leucine-rich repeats (LRRs), leucine zippers, nucleotide binding sites and kinase domains, reflecting related functions in signal transduction pathways (for review, see Ellis & Jones 1998). These recognition-type R genes commonly control high levels of resistance, yet they are vulnerable to pathogen adaptation due to rapid shifts in the pathogen population. However, some major genes also are associated with quantitative resistance (Koch & Parlevliet 1991, Martin & Ellingboe 1976, Nelson 1972). Many R genes are complex, and carry multiple homologous genes (for review, see Ellis & Jones 1998, Ronald 1998). Mapping studies have shown that some putative quantitative trait loci (QTL) are localized in chromosomal regions that harbour major R genes (McMullen & Simcox 1995, Wang et al 1994), although it has not been determined whether some QTL are R gene family members.

The second class of genes contributing to host plant resistance are the DR genes, which are recognized based on their increased expression during plant defence (for review, see Dixon & Harrison 1990). These genes are downstream from the specific host–pathogen recognition, and are considered less likely to be overcome by simple mutational changes in the pathogen. The proteins encoded by the *DR* genes include: (1) proteins that enhance cell wall structure (e.g. hydroxyproline-rich glycoproteins); (2) enzymes of secondary metabolism (e.g. phenylpropanoid pathway enzymes important in the synthesis of isoflavonoid and stilbene phytoalexins and lignin); and (3) enzymes which are implicated to be directly involved in the defence response, including chitinases, peroxidases, catalases, glucanases, sulfotransferases, and proteins that inactivate fungal ribosomes and bind chitin. Expression of *DR* genes can be induced by the interaction between the products of *R* genes and their corresponding pathogen avirulence genes, as in race-specific resistance, or by pathways involved in quantitative resistance.

Defence response genes in rice

Through map-based cloning and homology strategies, rice *R* genes are being cloned (Song et al 1995, Yoshimura et al 1998, Wang et al 1999) and genes for enzymes such as mitogen-activated protein kinases involved in the transduction of the *R–avr* gene signal are being identified (He et al 1999). *DR* genes in rice are those involved in the synthesis of secondary metabolites, including phenolic compounds, oxygenated fatty acid derivatives, lignin and phytoalexins (for review, see Leach et al 1996). Much effort has gone into identification of these *DR* genes (for review, see Wang & Leung 1998). As an example, the increased activity of at least three peroxidases is correlated with lignin deposition and resistance in rice to the bacterial blight pathogen, *Xanthomonas oryzae* pv. *oryzae* (Reimers et al 1992). Of five highly related peroxidase genes cloned from rice, three were induced during resistant responses (Chittoor et al 1997, E. Hilaire & J. Leach, unpublished results).

   In addition to the enzymes involved in biosynthetic pathways, enzymes that likely have a direct role in resistance also are induced in the rice defence response (Huang et al 1991, Xu et al 1996, Zhu & Lamb 1991). For example, enhanced expression of class I chitinases, which are implicated in hydrolysis of fungal pathogen cell walls, were observed in rice interactions with the blast fungus (Anuratha et al 1996, Xu et al 1996, Zhu & Lamb 1991), and *X. oryzae* pv. *oryzae* (J. Chittoor, J. Leach & F. White, unpublished results). Transgenic rice over-expressing chitinases exhibit higher levels of resistance to the blast and sheath blight fungi (Lin et al 1995), suggesting that these enzymes contribute to resistance.

The phenotype gap: mutational analysis to understand quantitative resistance

Although several lines of correlative biochemical, physiological, and gene-over-expression data are consistent with the putative roles of *DR* genes, there remains a large gap in our understanding of the functions of each of these genes. Glazebrook and colleagues (1997) approached this problem in *Arabidopsis* by isolating a series of phytoalexin (camalexin)-deficient (*pad*) mutants. Of the five *pad* mutants analysed, some (*pad1*, *pad2* and *pad4*) but not all, showed enhanced susceptibility when inoculated with a compatible bacterial pathogen, *Pseudomonas syringae* pv. *maculicola*. While single mutations in *PAD1*, *PAD2* and *PAD3* have little effect on resistance to a fungal pathogen, *Peronospora parasitica*, the double mutants show dramatically enhanced susceptibility. These studies in *Arabidopsis* illustrate an important point, i.e. depending on the position of the *DR* genes in the defence pathway(s), the genetic effects may vary according to different pathogen challenges. Through an analysis of single and double mutants, one can determine the pathogen specificity as well as functional redundancy of these *DR* genes.

Mutational analyses of *DR* genes, such as those described above for *Arabidopsis*, are so far limited in monocots (for review, see Piffanelli et al 1999), in general, and in particular, in rice. Using a mutational analysis, our goal is to narrow the gap in our understanding of what *DR* genes function and how the products of those genes function in quantitative and qualitative resistance.

Mutant production and screening for disease resistant mutants in rice

Mutants provide the biological variation essential for assigning function to the large amount of sequence information from various genome sequencing projects. Several laboratories around the world have begun producing various types of rice mutants as part of their functional genomics programs. For example, rice researchers are using heterologous transposons, gene and enhancer trap vectors as mutagenic agents for rice (An et al 2000, Chin et al 2000, Enoki et al 1999). The enhancer and gene trap insertion lines are advantageous not only for gene tagging but also for identifying sequences that affect growth stage or tissue-specific gene expression. The insertion lines can be examined at the heterozygous state, providing an opportunity to examine the role of essential genes, which would not be possible if homozygous mutations were lethal.

By exploiting the property of a retrotransposon (Tos17) that can be activated by tissue culture, Hirochika and co-workers have produced about 30 000 independent R2 lines that exhibit a large amount of variation (Hirochika et al 1996, Akio et al 2000). Each line contains an average of 5.5 Tos17 insertions; thus, the mutants contain about 165 000 insertion sites, representing a powerful tool for recovering tagged sequences and conducting reverse genetics. Another advantage of these retrotransposon lines is that they are non-transgenic and can be grown widely in the field. Replicated field observations are essential to identify subtle changes in phenotypes. Because of the need for tissue culture and transformation, however, all the transformation-mediated mutant collections are made in japonica rice; unfortunately, the indica type constitutes the bulk ($>90\%$) of rice grown in the tropical and subtropical environments.

To complement the insertion strategy, a large collection of mutants was produced at IRRI by using chemical mutagens (diepoxybutane, DEB) and irradiation (fast neutron, FN, and gamma ray, GR) in an indica rice variety IR64 that is widely grown in tropical Asia. To analyse $Xa21$-dependent signalling pathways, we made a collection of DEB and FN mutants in another indica variety, IRBB21, which contains the $Xa21$ bacterial blight resistance gene locus (G.L. Wang, P. Ronald & H. Leung, unpublished data). Our goal is to produce a range of deletion sizes such that genomic changes can be detected by reverse genetics as described by Westlund et al (1999) and Winzeler et al (1999).

Identification and characterization of mutants
in the Xa21-mediated resistance pathway

The bacterial blight resistance gene $Xa21$ confers resistance to many races of $X.$ *oryzae* pv. *oryzae* and encodes a receptor-like protein kinase (Song et al 1995, Wang et al 1996). To identify the genes required for the function of $Xa21$ resistance to $X.$ *oryzae* pv. *oryzae*, we screened families from IRBB21 ($Xa21$ donor) with DEB (3500 families) and FN (1000 families) for susceptibility to $X.$ *oryzae* pv. *oryzae* strain PXO99. We identified 33 plants with susceptible lesions (>8 cm) in 28 families, from which 11 M3 lines were confirmed to be highly susceptible (>20 cm) to $X.$ *oryzae* pv. *oryzae*, and 22 lines were found to be partially susceptible (6–20 cm).

Both PCR and Southern hybridization methods were used to detect deletions or re-arrangements in the $Xa21$ coding region. Some mutants had lost both the LRRs and kinase DNA fragments. Other mutants contained a fragment larger than the expected $Xa21$ LRR fragment (3 vs. 2 kb), suggesting that a re-arrangement had occurred. In these mutants, at least four $Xa21$ LRR-hybridizing bands were missing compared with the pattern in IRBB21. When the $Xa21$ kinase fragment was used as a probe, mutants with deletions in the LRR region showed only one hybridizing band. All mutants with deletions in the $Xa21$ gene cluster were highly susceptible to $X.$ *oryzae* pv. *oryzae*. In contrast, mutants with partial resistance to $X.$ *oryzae* pv. *oryzae* had no apparent deletions in the $Xa21$ locus as detected by both PCR and Southern analysis methods. Since mutants without deletion in the $Xa21$ locus were partially resistant to $X.$ *oryzae* pv. *oryzae*, the genes required for $Xa21$ resistance are either redundant or components of multiple pathways as described for disease resistance pathways of other plants (Innes 1998).

TABLE 1   Rice (IR64) mutant collection produced by chemical and irradiation mutagenesis[a]

Mutagen	Treatment	M_0 survival	M_1	M_2	M_3
Diepoxybutane	0.006%	60%	5420	3209	19 000
Fast neutron	33 Gy	90%	5400	2808	12 000
Gamma ray	250 Gy	95%	7350	5148	10 000

With header: *No. of plants/families advanced* spanning M_1, M_2, M_3.

[a]Both fast neutron and gamma ray treatments were performed by Drs P. Donini and F. Zapata at the International Atomic Energy Commission, Vienna, Austria.

Isolation of disease response mutants in the IR64 mutant collection

IR64 carries many valuable agronomic traits related to yield, plant architecture, grain quality, and tolerance to biotic and abiotic stresses. Thus, creating mutations in this elite genetic background can facilitate the detection of phenotypic changes in important agronomic traits. Our target is to produce 40 000 independent deletion lines to give a high probability of detecting a mutation in most genes, with the exception of homozygous lethals. Mutagenesis with DEB, FN and GR gives mutation rates of 0.3–1% in disease response traits (Table 1), a result similar to that observed in other plant species (Okubara et al 1994).

From a screen of about 10 000 independent M2 lines, we have thus far genetically confirmed about 30 mutants with altered response to one or two pathogens (Table 2). Not all mutants in the M2 population have been analysed. However, based on our results so far, we conservatively estimate a 0.3% detection frequency in defence response mutations. Additional mutant categories will be discovered by expanding the mutant screen. Using the categories of mutants shown in Table 2, we have begun to systematically produce double mutants which will be essential for dissecting the DR pathways through a combination of phenotypic and gene expression analysis.

To identify loss-of-resistance mutants, the M2 population was first screened with a pool of four rice blast isolates that are avirulent to IR64. Gain-of-resistance was screened separately using a single isolate (Ca89) virulent on IR64. Candidate mutants were then screened against bacterial blight strains. As has been observed in other plant species (Innes 1998), a common class of mutants found contain recessive mutations resulting in the loss of resistance to a single pathogen race. Loss of resistance to multiple races can be conditioned by either recessive or

TABLE 2   Categories of disease response mutants of IR64 against blast and bacterial blight

		Blast		Blight		
		Number of mutants [b]				
Mutation	Morphology[a]	*Single race*	*Multiple race*	*Single race*	*Multiple race*	*Both diseases*
Loss of resistance	Normal	—	2	2	1	—
Gain of resistance	Normal	2	3	2	—	3
	Abnormal	2	—	2	—	2
Lesion mimics	Normal	—	—	—	—	—
	Abnormal	—	—	1	—	3

[a]Abnormal morphology includes dwarfing, stunting and reduced tillering.
[b]Numbers refer to mutants confirmed by M3 segregation data. Four *M. grisea* isolates representing diverse genetic groups were used to screen for susceptibility to blast. *X. oryzae* pv. *oryzae* isolate PXO61 (race 1) was used to screen for a change from resistance to susceptibility to bacterial blight. PXO99 (race 6) was used to screen for a change from susceptible to resistance.

dominant mutations. Mutants with susceptibility to all blast isolates and to both blast and blight diseases have not yet been found.

Four mutants with a gain-of-resistance to a virulent blast isolate Ca89 were identified. Half (50%) of these mutants are morphologically abnormal (stunting, dwarfing, reduced tillering). The morphologically normal mutants from this group are not resistant to bacterial blight. However, gain in resistance to both blast and blight was observed in at least two developmentally abnormal mutants. In *Arabidopsis*, some gain-of-resistance mutants were also morphologically abnormal (Rate et al 1999).

Lesion mimic mutations to dissect disease resistance pathways in rice

To investigate the role of the hypersensitive response (HR) and to dissect the resistance pathways, lesion mimic mutants have been characterized and the genes responsible for the mutations have been cloned in *Arabidopsis*, barley and maize (e.g. Greenberg et al 1994, Dangl et al 1996, Dietrich et al 1997, Gray et al 1997, Büschges et al 1997). In rice, we have characterized eight previously identified spotted leaf (*spl*) mutants (*spl1*, *spl2*, *spl3*, *spl4*, *spl5*, *spl6*, *spl9* and *spl11*). Mutants *spl1–spl9* are near-isogenic lines in the IR36 background (BC3F3); *spl11* is an EMS mutant from cultivar IR68 (Singh et al 1995). These mutants show different types of mimic lesions on leaves at different growth stages.

Since most *spl* mutants show the characteristics of a resistant reaction following blast infection, the mutations may occur in the pathway leading to resistance. To test this, we assessed the reaction of the various lesion mimic mutants to 10 different rice blast isolates. Three mutants (*spl1*, *spl5* and *spl9*) showed smaller or no lesions in response to all 10 isolates as compared with the susceptible recurrent parent IR36. After eliminating the endogenous blast resistance genes from the IR68 background, plants containing *spl11* were also found to be resistant to all 10 blast isolates. Interestingly, *spl11* (but not the other mutants) also conferred a high level of resistance to four different bacterial blight strains. Thus mutations in some of these rice lesion mimic mutants may be involved in the resistance pathways to two different rice pathogens.

Assigning phenotypes to candidate defence response genes

While forward genetics allows identification of mutations with detectable phenotypes, it does not permit the isolation of mutations in defence genes that produce subtle changes or accumulative effects. With the availability of candidate *DR* genes, the rice deletion mutants can be used to assign mutations to these candidate genes by a reverse genetic approach. Collections of completely or partially sequenced cDNAs (ESTs) have been generated (Rounsley et al 1998) and, based upon structural similarity, some randomly sequenced cDNAs are suspected to code for enzymes involved in disease defence. For example, G.-H. Miao (DuPont, USA) has identified about 800 ESTs with significant similarity to the cloned resistance and defence response genes. In addition, through database mining, we have established a collection of over 150 putative disease defence response and resistance genes, the KSU Defense Gene Collection (KSU-DGC). These genes and their sequences can be used (1) in screens to identify rice mutants with deletions or other structural rearrangements in *DR* genes and (2) to characterize the expression profiles of individual mutants. As an example, using two defence-related genes as probes (an NBS-LRR sequence and a MAP kinase gene), we have detected genomic changes in two blast-susceptible rice deletion mutants. The hybridization patterns suggest deletions and chromosomal re-arrangements occurred in these mutants. As more EST and candidate genes are available, we expect an increasing probability of associating mutant phenotypes with DNA sequences.

DNA array analysis

By examining the gene expression profiles of these mutants under different pathogen challenges, genes that are up- or down-regulated by the mutation can be identified. In collaboration with G.-H. Miao (DuPont), G. L. Wang (unpublished results) used microarrays to detect differential gene expression in near isogenic lines carrying single resistance genes (*Pi1*, *Pi2*, *xa5* and *Xa21*) infected with blast and bacterial blight pathogens. Of the 1536 EST clones used as targets for microarrays, half (768) were homologous to known R genes and host defence response genes in plants, while the rest were randomly picked from a normalized rice cDNA library. An initial screen showed that 258 ESTs were induced and 571 ESTs were repressed in the resistance response of all four R genes at all time-points after inoculation with bacterial blight and blast isolates, respectively. These results, though preliminary, indicate the power of high-throughput gene expression analysis towards understanding defence mechanisms in different rice genotypes.

Summary

The dissection of disease response pathways has significance in understanding the basic biology of host–pathogen interactions and practical disease control. Using the variability provided by a rich collection of rice mutants, structural sequence information and high-throughput genomic tools can be applied to unravel the biological function of genes in the defence response pathway. The pool of candidate genes identified can be used for designing rice genotypes with broad-spectrum resistance to multiple diseases.

References

Akio M, Murata K, Tanaka K et al 2000 Systematic analysis of mutations of rice induced by retrotransposon Tos17 insertion. In: Abstracts of plant and animal genome VIII, San Diego, CA, January 2000, p 99 (*http://www.intl-pag.org/pag/*)

An G, Jeon JS, Lee SC, et al 2000 T-DNA tagging of rice genome. In: Abstracts of plant and animal genome VIII, San Diego, CA, January 2000, p 100 (*http://www.intl-pag.org/pag/*)

Anuratha CS, Zen KC, Cole KC, Mew T, Muthukrishnan S 1996 Induction of chitinases and β glucanases in *Rhizoctonia solani*-infected rice plants: isolation of an infection-related chitinase cDNA clone. Physiol Plant 97:39–46

Buschges R, Hollricher K, Panstruga R et al 1997 The barley *mlo* gene: a novel control element of plant pathogen resistance. Cell 88:695–705

Chin H, Choe M, Oh B et al 2000 Molecular analysis of rice plants harboring an Ac/Ds transposable element-mediated gene trapping system. In: Abstracts of plant and animal genome VIII, San Diego, CA, January 2000, p 100 (*http://www.intl-pag.org/pag/*)

Chittoor JM, Leach JE, White FF 1997 Differential induction of a peroxidase gene family during infection of rice by *Xanthomonas oryzae* pv. *oryzae*. Mol Plant-Microbe Interact 10:861–871

Dangl JL, Dietrich RA, Richberg MH 1996 Death don't have no mercy: cell death programs in plant–microbe interactions. Plant Cell 8:1793–1807

Dietrich R, Richberg MH, Schmid R, Dean C, Dangl JL 1997 A novel zinc finger protein is encoded by the Arabidopsis LSD1 gene and functions as a negative regulator of plant cell death. Cell 88:685–694

Dixon RA, Harrison MJ 1990 Activation, structure, and organization of genes involved in microbial defense in plants. Adv Genet 28:165–234

Ellis J, Jones D 1998 Structure and function of proteins controlling strain-specific pathogen resistance in plants. Curr Opin Plant Biol 1:288–293

Enoki H, Izawa T, Kawahara M et al 1999 Ac as a tool for the functional genomics of rice. Plant J 19:605–613

Glazebrook J, Rogers EE, Ausubel FM 1997 Use of *Arabidopsis* for genetic dissection of plant defense responses. Annu Rev Genet 31:547–569

Gray J, Close PS, Briggs SP, Johal GS 1997 A novel suppressor of cell death in plants encoded by the *Lls1* gene of maize. Cell 89:25–31

Greenberg JT, Guo A, Klessig DF, Ausubel FM 1994 Programmed cell death in plants: a pathogen-triggered response activated coordinately with multiple defense functions. Cell 77:551–564

He C, Fong SHT, Yang D, Wang GL 1999 BWMK1, a novel MAP kinase induced by fungal infection and mechanical wounding in rice. Mol Plant-Microbe Interact 12:1064–1073

Hirochika H, Sugimoto K, Otsuki Y, Tsugawa H, Kanda M 1996 Retrotransposons of rice involved in mutations induced by tissue culture. Proc Natl Acad Sci USA 93:7783–7788

Huang J, Wen L, Swegle M et al 1991 Nucleotide sequence of a rice genomic clone that encodes a class I endochitinase. Plant Mol Biol 16:479–480

Innes RW 1998 Genetic dissection of R gene signal transduction pathway. Curr Opin Plant Biol 1:299–304

Koch M, Parlevliet JE 1991 Residual effects of the *Xa4* resistance gene in three rice cultivars when exposed to a virulent isolate of *Xanthomonas campestris* pv. *oryzae*. Euphytica 55:187–193

Leach JE, Young SA, Chittoor JM, Zhu W, White F 1996 Induction of defense responses in rice. In: Mills D, Kunoh H, Keen N, Mayama S (eds) Molecular aspects of pathogenicity and host resistance: requirements for signal transduction, Academic Press, New York, p 115–128

Lin W, Anuratha CS, Datta K, Potrykus I, Muthukrishnan S, Datta SK 1995 Genetic engineering of rice for resistance to sheath blight. Bio-Technology 13:686–691

Martin TJ, Ellingboe AH 1976 Differences between compatible parasite/host genotypes involving the Pm4 locus of wheat and the corresponding genes in *Erysiphe graminis* f. sp. *tritici*. Phytopathology 66:1435–1438

McMullen MD, Simcox K 1995 Genomic organization of disease and insect resistance genes in maize. Mol Plant-Microbe Interact 8:811–815

Nelson RR 1972 Genetics of horizontal resistance to plant diseases. Annu Rev Phytopathol 16:359–378

Okubara PA, Anderson PA, Ochoa OE, Michelmore RW 1994 Mutants of downy mildew resistance in *Lactuca sativa* (lettuce). Genetics 137:867–874

Piffanelli P, Devoto A, Schulze-Lefert P 1999 Defense signaling pathways in cereals. Curr Opin Plant Biol 2:295–300

Rate DN, Cuenca JV, Bowman GR, Guttman DS, Greenberg JT 1999 The gain-of-function *Arabidopsis acd6* mutant reveals novel regulation and function of the salicylic acid signaling pathway in controlling cell death, defenses, and cell growth. Plant Cell 11:1695–1708

Reimers P J, Guo A, Leach JE 1992 Increased activity of a cationic peroxidase associated with an incompatible interaction between *Xanthomonas oryzae* pv. *oryzae* and rice (*Oryza sativa*). Plant Physiol 99:1044–1050

Ronald PC 1998 Resistance gene evolution. Curr Opin Plant Biol 1:294–298

Rounsley S, Lin X, Ketchum KS 1998 Large-scale sequencing of plant genomes. Curr Opin Plant Biol 1:136–141

Singh K, Multani DS, Khush G 1995 A new spotted leaf mutant in rice. Rice Genet Newsl 12:192–193

Song WY, Wang GL, Chen LL, et al 1995 A receptor kinase-like protein encoded by the rice disease resistance gene, *Xa21*. Science 270:1804–1806

Wang GL, Leung H 1998 Molecular biology of host–pathogen interactions in rice diseases. In: Shimamoto K (ed) Molecular biology of rice. Springer-Verlag, Tokyo, p 232–241

Wang GL, Mackill DJ, Bonman JM, McCouch SR, Nelson RJ 1994 RFLP mapping of genes conferring complete and partial resistance to blast resistance in a durably resistant rice cultivar. Genetics 136:1421–1434

Wang GL, Song WY, Ruan DL, Sideris S, Ronald PC 1996 The cloned gene, *Xa21*, confers resistance to multiple *Xanthomonas oryzae* pv. *oryzae* isolates in transgenic plants. Mol Plant-Microbe Interact 9:850–855

Wang ZX, Yano M, Yamanouchi U et al 1999 The *Pib* for rice blast resistance belongs to the nucleotide binding and leucine-rich repeat class of plant disease resistance genes. Plant J 19:55–64

Westlund B, Parry D, Clover R, Basson M, Johnson CD 1999 Reverse genetic analysis of *Caenorhabditis elegans* presenilins reveals redundant but unequal roles for *sel-12* and *hop-1* in Notch-pathway signaling. Proc Natl Acad Sci USA 96:2497–2502

Winzeler EA, Shoemaker DD, Astromoff A et al 1999 Functional characterization of the *S. cerevisiae* genome by gene deletion and parallel analysis. Science 285:901–906

Xu Y, Zhu Q, Panbangred W, Shirasu K, Lamb C 1996 Regulation, expression and function of a new basic chitinase gene in rice (*Oryza sativa* L.). Plant Mol Biol 30:387–401

Yoshimura S, Yamanouchi U, Katayose Y et al 1998 Expression of *Xa1*, a bacterial blight-resistance gene in rice, is induced by bacterial inoculation. Proc Natl Acad Sci USA 95:1663–1668

Zhu Q, Lamb CJ 1991 Isolation and characterization of a rice gene encoding a basic chitinase. Mol Gen Genet 226:289–296

DISCUSSION

Gale: Hei Leung, we began a discussion earlier about how many mutations you might expect to be in each of these lines. You said 10–20, and Jan Leach has just said 10. I would like to question this figure. Jan Leach showed two primers for which you got null results. I am guessing that these primers are 20 base pairs each, and being conservative in assuming that any base pair change would stop the primers from working. This is one ten-millionth of the genome. If you had two of these out of, and I'm guessing, say 1000 screened lines, and you were averagely lucky, you would need 20 000 mutations per line, not 10.

Leung: I agree that 10 is probably a conservative estimate. The way I calculate is not based so much on primer pairs, but rather on the frequency of mutation in one particular trait. If we take one locus, *Xa21*, we find that out of 4000 M2 lines, we get 30 mutants. This is roughly one-in-1000 recovery of a single locus mutation, which is consistent with most disease screens in *Arabidopsis*. I work backwards and figure that in order to see this one in 1000 mutation, the number of hits per M1 genome has a minimum of 10–20. It could be higher than this, but I don't think it will be in the 1000 range. For our purpose, of screening biological variation the more mutations the better for our stock as long as they are not lethal.

Gale: I'm not absolutely sure that you are better off with more, because you will get more false leads.

Salmeron: Wouldn't the number of mutations you get by treatment be some function of the total genome size? You said you were getting 60–90% survival. Wouldn't the number of physical mutations required to generate that vary depending on the genome size? In a complex target there is a lot of the genome which, if mutated, would have no effect.

Leung: The 90% survival that I mentioned earlier referred to the gamma ray hit. The diepoxybutane treatment gives about 60% M2 recovery. I have struggled with the numbers I gave for the mutation rates for some time, and I would welcome help from people with more experience in this.

Dong: It is surprising to me that the survival rate for the fast neutron mutagenesis is 90%.

Leung: That is the M1 survival rate, when the plants are in the heterozygous state. We only get 5% kill for the gamma rays, but with diepoxybutane we see 60% kill.

Ku: Why don't you use transposon tagging to generate these mutants? Is this related to the problem of matching the gene with the phenotype?

Leung: Had we started a few years earlier, we would probably have used transposon tagging. When we initiated this project, there were already a number of labs around the world who were already into the second year of the *Ac/Ds* transposon tagging or enhancer trap studies. For this reason our strategy is to fast-track the production of indica rice, which is more difficult to transform. In addition, using chemical or irradiation mutagenesis, we can grow the non-transgenic plants in the field and produce a lot of seed. Because these plants are non-transgenic, they can be grown everywhere. Ideally, we would like to have a collection such as the *Tos17* type retrotransposon population. We hope that through collaboration we can have access to this sort of resource. The deletion stock is something we want to move fast so that we can generate the biological variation and really get involved in the functional genomics programme. The ideal scenario is that within two or three years we will have both insertion and deletion lines running in parallel. For indica rice we will use chemical and radiation mutagenesis, and for japonica rice we will get insertion lines.

Salmeron: I have a question about the disease mutants you have so far identified. In general, how far have you taken these? Are they single loci?

Leung: For the category mutants that Jan Leach showed, these are M3 data. That is, we do know the inheritance because they segregate within and between lines. We currently have approximately 30 mutants for which the genetics have been confirmed. In total, if we consider them collectively as disease response mutants, we are recovering at about 0.3% since we have screened about 10 000 independent lines. For the gain of resistance mutants, many have abnormal morphology. We are not using this mutagenesis screen for plant breeding; we are using it for dissecting the pathways. We have not seen anything that has broad-spectrum susceptibility yet.

G.-L. Wang: We screened nine lesion mimic mutants. Three showed enhanced resistance to rice blast, and only one showed resistance to both pathogens. It will take more screening to get that kind of mutant in the IR64 mutant collections.

Dong: With deletion mutants you can still consider using genomic subtraction to clone the gene. If you can get 5–10 alleles of the same genetic locus, then it may not be a bad idea to try genomic subtraction.

G.-L. Wang: That is one way. Another method in the near future might be using the bacterial artificial chromosome (BAC) contigs from the genome project.

Parker: As a complementary approach to targeting particular genes for function in resistance pathways, has anyone thought of adopting the virus-induced gene silencing strategy in rice? Here, a gene is expressed on a virus and then this co-suppresses the homologous gene or gene family in the host plant.

G.-L. Wang: I think there are many labs interested in making this technique work in monocots. For rice, the problem is that we don't have a vector system.

Bouis: I am an economist by training and I don't understand a lot of the technical issues discussed here. Later in this symposium we will be discussing breeding for nutrition, using either conventional methods or biotechnology. There are many other ways of trying to address nutrition, either by giving supplements or by fortifying foods, for example. Thus the justification for following plant breeding is that it turns out to be very much more cost-effective. I am trying to get some perspective on this discussion about the work on plant diseases. There must be other ways than biotechnology for trying to combat disease. Is this research on plant resistance because there is no other way to get disease resistance? Are there ways you can get resistance through conventional breeding? What is the motivation behind this research?

Dong: It is not possible to say that conventional breeding is cheaper, because there are not enough examples of genetic engineering to act as a comparison. The start-up costs of any venture tend to be large, and the same is true with crop biotechnology, but this will become cheaper.

Leach: I don't think there is any reason to believe that the disease resistance we are working towards can only be introduced by genetic engineering. What we are trying to understand is what makes plants durably resistant, and unravelling the pathways that lead to this resistance. Modification of these pathways by genetic engineering may be one approach, but we can possibly direct breeders also by understanding the pathways.

Leung: What we are discussing here is probably not an alternative agriculture. I think biotechnology is mainstream agriculture: it is the application of molecular science to facilitate the understanding of various biological mechanisms and thus to allow more precision in breeding programmes. The breeding programme at IRRI is integrated with multiple techniques: there is no division between biotech and conventional breeding. The issue of transgenic crops is a distinct one because of genetically modified organisms (GMOs). Molecular science is not necessarily equivalent to transgenic technology; it is the application of science to understand gene function and germplasm utilization. We should do our best to uncouple GMOs and biotechnology.

Khush: It is really a continuum. Conventional breeding has been used to produce disease-resistant plants, but there are some things that cannot be done through conventional breeding yet which can be achieved through biotechnology. Let's leave aside the issue of GMOs, which is currently problematic. There are techniques such as molecular marker-aided selection, which can be used to combine several genes for resistance to bacterial blight, for example. The purpose of this whole molecular analysis is to give an understanding of the mechanisms of resistance, and develop techniques for producing more durable resistance.

Presting: I was sitting here thinking about the mutagenesis approach. It occurred to me that mutagenesis was very popular in the 1960s. With the rice genome about to be sequenced, a lot of ESTs are already available, so how about doing a more directed approach using single or multiple ESTs for co-suppression analysis? In this way we don't have to worry about looking at where the deletions are or what the genes are. Rather, you can correlate the phenotype with the gene that has been co-suppressed straight away. We now have something we didn't have in the 1960s: genomic information.

Leach: It can work, and sometimes it doesn't work. For example, with large gene families such as phospholipase D or peroxidase, there is frequently compensation between family members. If there is compensation from another family member that is slightly related, it is difficult to come up with conclusive evidence. From my perspective, a knockout is an ideal tool.

Li: Recently, at IRRI we have developed transgenic lines for the four bacterial blight resistance genes. We find interesting differences between the recessive and dominant resistance genes. The dominant genes tend to have a strong residual effect. The recessive genes don't, but the recessive genes tend to have stronger epistatic effects when interacting with each other and with the dominant genes. What do these differences mean in terms of their roles in terms of plant defensive pathways?

Leach: All resistance genes are not created equal, and their interactions with the pathogens are not equal. We did a field study here in the Philippines that clearly shows differences in potential durability of different bacterial blight resistance genes. For example, *Xa7*, which isn't a rapid-type hypersensitive response, would be an extremely durable resistance gene on the basis of our predictions and field studies. This is because the loss of the corresponding pathogen *avirulence (avr)* gene really costs the pathogen in terms of not only aggressiveness to the plant but also fitness. In contrast, *Xa10*, which causes a strong and rapid hypersensitive response, probably won't be durable on the basis of field studies and also the fact that when the pathogen factor (*avr* gene) is knocked out, it doesn't cost the pathogen anything in terms of aggressiveness or fitness. We don't have field studies on *xa5*, which is a recessive gene, but based on lab studies on the knockouts of the *avr* genes, I predict that this will be a relatively durable resistance gene: when the pathogen *avr* gene is knocked out, it costs the pathogen. We have correlated our lab and field studies to try to understand and predict durability of resistance genes. Thus, the resistance genes are not all created equal, and we need to be able to predict which ones are the best.

Li: I need to emphasize that *xa5* is not a completely recessive gene. The lesion length of the F1 plants between near isogenic pairs is between the recessive and resistant NIL parents. Only in one case, in race 4, is it completely recessive.

Nevill: Can you take any of the principles for signal transduction in disease resistance and apply these to resistance factors for insects?

Leung: In insect interactions there are fewer intimate relationships found as in pathogen–host plant interactions, except for well defined systems like the gall midge or Hessian fly, with a gene-for-gene relationship.

Khush: A gene-for-gene relationship has been found in several insects, such as the Hessian fly in wheat, the brown plant hopper and gall midge. But I agree that the relationship is not as intimate as with the pathogens.

Parker: I would perhaps counter that with the example of the *mi* gene in tomato. This gene confers dual specificity resistance both to a nematode and an aphid. I think there will be common mechanisms. There are also a number of studies in plants showing an antagonism between pathways that involve salicylic acid in pathogen resistance, and pathways that involve jasmonic acid in the insect wound response. I think there is going to be a lot learned in the next few years about the cross-talk of these various responses.

Khush: Is there anyone working on the molecular biology of insect resistance?

Parker: There are a growing number of groups, for example the Institute of Chemical Ecology in Jena, Gemany.

Salmeron: An explanation for why there is much more work on diseases rather than insect pests is simply that it is easier to work with diseases in the lab.

Khush: This is why there has been less breeding for insect resistance. However, we have had some success in breeding for insect resistance at IRRI.

Breeding for nutritional characteristics in cereals

Robin D. Graham and Glenn Gregorio*

*Department of Plant Science, University of Adelaide, Glen Osmond, 5064, South Australia, Australia, and *Plant Breeding, Genetics and Biochemistry Division, International Rice Research Institute, Los Baños, Laguna, The Philippines*

Abstract. Extensive genetic variation within large species such as the major cereals can be confidently expected for any new trait of interest. This has now been extensively demonstrated for the nutrient content of cereal grains that is of interest under deficient conditions both to human nutritionists and to cereal agronomists. As cereals are eaten in large quantity by practically everyone, they are the ideal vehicles for changing the balance of nutrient intake of the whole human population. Doing so appears to be necessary as the World Health Organization has identified deficient micronutrient intake in well over half of all people globally, notably women and children. Of major concern are iron, zinc, selenium iodine, calcium and vitamin A-related carotenoids. Our results show that for any staple so far studied, the intake of iron, calcium and zinc from cereals can be doubled, and the content/intake of essential carotenoids can be increased by much greater factors. To prove to rigid scientific standards that greater intake results in greater absorption and measurable health benefits is quite difficult, but it is currently being pursued in various ways. This proof of bioavailability is all that impedes implementation in breeding programs.

2001 Rice biotechnology: improving yield, stress tolerance and grain quality. Wiley, Chichester (Novartis Foundation Symposium 236) p 205–218

The web site of the World Health Organization (WHO) gives the awesome statistic that over half of the world's people are deficient in iron, that is, more than 3 billion people; two billion are at risk of iodine deficiency, and 230 million children are vitamin A deficient. While no easy test for zinc deficiency exists and so there are no statistics, specialists in zinc nutrition consider it may be as widespread as iron deficiency. Supplementation with these nutrients by injection or pills can be effective and economically viable. However, almost without exception where these programs in areas of high incidence of micronutrient deficiencies have been evaluated, they have failed to reach all people at risk for lack of adequate infrastructure and education. We have concluded that for developing countries, a food-based system of delivering all nutrients in adequate amounts is the sustainable solution. These same nutrient deficiencies are also major public health issues for developed countries, though the figures are not nearly as alarming; the problem is qualitatively the same and we argue that the sustainable solution is also the same.

A new paradigm for agriculture

The current paradigm of agriculture, the sustainable productivity paradigm, takes account of the need for increasing production of food while protecting the natural resource base of that production, but it does not address the nutritional quality of the food its systems develop. The need for change has been brought into sharp focus by the statistics on nutrient deficiencies of the WHO and the World Bank (Anonymous 1992, 1994) in the last few years.

Most economists believe that micronutrient malnutrition is best solved by increasing the income of those afflicted, allowing those families to buy more food, diversify their diets, and thereby obtain adequate nutrient balance. Unfortunately, this has not always worked over the past 30 years in many developing countries. For example, as a result of the 'green revolution' caloric intake and incomes rose substantially globally, but iron deficiency dramatically increased at the same time, even in those societies where appreciable gains were made in both income and in staple food production (Anonymous 1994). Apparently, means of achieving balanced nutrition need to be developed other than by relying on increasing income and cereal production. We submit that improved micronutrient output of agricultural systems is a necessary prerequisite for eliminating micronutrient malnutrition in a sustainable fashion.

A change in priorities from sustainable production of 'food' to sustainable delivery of nutrients in balance through health-focused food systems is best embodied in a new paradigm for agriculture: sustainable, productive, nutritious food systems.

Nutrient interactions involving iron, zinc and vitamin A

Iron, zinc and vitamin A are the three micronutrients of major concern about which we have a sizeable body of agricultural information. Much less is known of the genotypic variation of iodine and selenium, two more elements that are widely deficient. The density of iron, zinc and pro-vitamin A can be increased in staples by plant breeding but the size of the increase that is necessary is not clear as no feeding trials comparing varieties have been done and the effects of interactions in the human body are only just emerging. An interaction is recognized when the response of an individual to one nutrient is not constant but varies depending on the level of another nutrient. In terms of nutrition and health, the essential roles (main effects) of these three nutrients are well known and do not need to be repeated here, but recent literature indicates strong, positive synergies between these nutrients in deficient individuals that may in part determine the breeding objectives we must achieve.

Zinc–vitamin A interactions

Evidence of interactions between vitamin A and zinc has emerged in the late 20th century (see reviews by Christian & West 1998, Solomons & Russell 1980). In their summary figure, Christian & West (1998) indicate a role of zinc in synthesis of retinol binding protein (RBP) which is involved in releasing vitamin A from the liver, and in increasing lymphatic absorption of retinol and its inter- and intracellular transport. To return the favour, vitamin A promotes the synthesis of a zinc-dependent binding protein and thereby the absorption and lymphatic transport of zinc. The interaction of delivering these two essential nutrients together to patients deficient in both was shown by Udomkesmalee et al (1992) who concluded that the dual treatment was so effective that supplementation with as little as twice the recommended daily allowance of both together was sufficient to normalize blood indices of the cohort. Plant breeding could also double the normal daily dose of these nutrients compared to that delivered by today's staples.

Iron–zinc interactions

Both synergistic and antagonistic interactions occur; the competition of Fe^{2+} on Zn^{2+} for the bond with plasma transferrin is an antagonistic interaction that is well documented (Georgievskii et al 1982). However, this effect is less likely to be significant if the subject is deficient in these nutrients and less likely still when iron and zinc are given as food than as soluble supplements (Whittiker 1998). Iron and zinc are considered to be synergistic in promoting work efficiency of muscle, in promoting cognitive ability and in depositing calcium in bone. Indeed, anaemia is one of the symptoms of zinc deficiency (Prasad 1984), implying a synergistic effect of zinc on iron absorption and transport or on haematopoiesis itself. Such synergistic effects between iron and zinc, especially if shown to act on gastrointestinal absorption of each other from cereal diets, strongly indicate the breeding for staples dense in both iron and zinc in order to address iron deficiency anaemia effectively.

Vitamin A–iron interactions

A seminal paper by Hodges et al (1978) aroused concern about the importance of this interaction. They showed that humans kept on a vitamin A-deficient diet became anaemic despite adequate iron intake. The anaemia responded to vitamin A and not to iron supplement. Several supplementation studies established that dual supplements were more effective than either alone (Muhilal et al 1988, Panth et al 1990). Interest in this interaction has been kindled by the recent paper by García-Casal et al (1998) who showed that just 500 i.u. of vitamin A or β-carotene added to a 0.1 kg meal of cereal (wheat, rice or maize) doubled the iron absorption from the gut of human subjects in Venezuela. It appears that in the presence of high levels of phytate and tannins in the diet, vitamin A or β-carotene will enhance the bioavailability of iron in humans, but in the absence of these antinutrients, no enhancement was found (R. M. Welch & D. L. Garvin, personal communication). The data suggest a role for provitamin A carotenoids in absorption, transport and function of iron, and vice versa.

A putative second-order interaction

Although the study may be technically difficult, a three-way interaction in which the response to one nutrient depends on the levels of the other two nutrients can be predicted. Given the synergies in all three first-order interactions, an individual deficient in all three nutrients can be expected to respond dramatically to relatively small supplements of the three nutrients given together, until normal homeostasis is reached. The importance of these interactions is underlined by the reports that vitamin A deficiencies are appearing in countries in which it had not previously been a problem (Darnton-Hill et al 1992). We conclude that increasing the iron and zinc by not more than double and increasing the carotene content to that which would supply at least half the recommended daily allowance (RDA), in the one variety of staple, would likely eliminate all three deficiencies in most individuals at risk.

Increasing micronutrients in food systems

Two key agricultural strategies for increasing micronutrients in food systems are fertilizer use and plant breeding.

Fertilizer

Fertilizer technology and use is widely understood and appreciated in modern agriculture so it can be a major vehicle for change in plant mineral content and food quality. The density of several micronutrients can be usefully enhanced by application of the appropriate mineral forms (Allaway 1986, House & Welch 1989): these are zinc, iodine, selenium, copper and nickel (Table 1).

TABLE 1 Zinc concentrations (mg/kg, dry wt) in the grain of wheat cultivars grown on zinc-deficient soil in South Australia, with and without zinc fertilizer added to the soil at sowing

	Zinc concentration (mg/kg)	
Cultivar	−Zn	+Zn
Excalibur	7.6	25.3
Warigal	13.0	26.0
Kamilaroi	8.8	23.4
LSD(0.05)	3.4	

From Graham et al (1992).

However, because of rapid oxidation in soil and because of low mobility in phloem, soluble fertilizers of iron are ineffective in increasing the iron concentration in plants, especially in the seeds that develop months after application. Foliar applications are not much better. All the vitamins of plant origin are synthesised *de novo* by the plant and are not a consideration as fertilizers. Thus, for many of the mineral nutrients of concern, fertilization is a useful strategy, while for iron and the vitamins, it is not (although adequate plant nutrition is a prerequisite for optimum vitamin biosynthesis in plants).

Plant breeding

The findings in rice are particularly encouraging for the plant breeding strategy. Iron density in rice varied from 6.3–24.4 mg kg^{-1} (all concentrations reported are on a dry weight basis) and zinc density from 15.3–58.4 mg kg^{-1}. A benchmark was established in that nearly all the widely grown 'green revolution' varieties were similar, about 12 and 22 mg kg^{-1} for iron and zinc respectively. The best lines discovered in the survey of the germplasm collection were therefore twice as high in iron and 1.6 times as high in zinc as the most widely grown varieties today. High iron, and to a lesser extent high zinc, concentrations were subsequently shown to be only weakly linked to the trait of aromaticity, but most aromatic rice types such as Milagrosa and Basmati are high in both iron and zinc, and as before, generally higher in most minerals (Table 2). Aroma in rice is dominated by a major gene whereas iron density is linked to a suite of four QTLs (Gregorio et al 2000). As in other crops, these micronutrient density traits have been combined with high yield.

In addition to genetic variation in nutrient density in rice grains, there are genotypic differences in the relative loss of iron and zinc in the milling process.

TABLE 2 Elevated concentrations of iron and zinc in aromatic rice varieties in comparison with non-aromatic types grown at the International Rice Research Institute, Los Baños in 1996

Aromatic Line ID	Fe mg/kg	Zn	Non-aromatic Line ID	Fe mg/kg	Zn
Set 1					
Basmati 370	16.3	34.4	IR8	12.3	17.3
Gaok	16.0	26.4	IR36	11.8	23.1
Azucena	18.2	29.3	IR74	11.2	24.0
Mean	16.8	30.0		11.7	21.5
Set 5					
Ganje Roozy	18.1	36.6	Bg 379-2	11.3	20.5
Banjaiman	18.1	33.3	BG1370	11.5	19.5
CT 7127	17.1	32.4	UPLRi 7	10.8	20.9
Lagrue	19.0	34.8	Tetep	10.7	24.1
Mean	18.1	34.0		11.1	21.3

Data of D. Senadhira, from Graham et al (1999).

Rice is the poorest in iron of all the cereals because iron is inherently low in rice and milling removes more than half of the total iron in the grain.

Similar genetic variation has been found for the other cereals (Graham et al 1999).

Carotenoids

The prospects for improving the provitamin A (carotene) density of staple plant foods are very good indeed. Yellow types are well known in all the usual white starchy plant foods. Although the genetics are complex and somewhat obscure, rapid breeding progress is nevertheless possible since genetic gain in carotene content can be visually estimated with sufficient accuracy (Simon 1992). Recently, carotenoids other than β-carotene have been shown to be present in the eye and important to preventing macular degeneration (Khachik et al 1997). Such carotenoids are contained in pasta wheats and were common in older wheat varieties (Rosser et al 1999). This century, wheat breeding has been focused on wheats producing white flour, driven by market demand, and the traits for high grain carotenoid content have been largely eliminated from modern germplasm. By re-educating consumers in western countries, these types could be brought back into breeding programs in all countries. High carotene yellow maize has eliminated vitamin A deficiency in pigs housed in winter (Brunson & Quackenbush 1962). For β-carotene in maize and cassava, the range is much greater than that for iron (Brunson & Quackenbush 1962, Iglesias et al 1997).

Seedling vigour

The micronutrient density of seeds is important not only for human nutrition, but also for the nutrition of the seedling in the next generation. A more vigorous crop is established by seeds with a high density of nutrients, including micronutrients (Welch 1986, Rengel & Graham 1995, Moussavi-Nik et al 1997), that confer better resistance to stress and disease, leading eventually to higher yield. These effects are most pronounced in micronutrient-deficient soils (Weir & Hudson 1966, Yilmaz et al 1997). Thus, micronutrient-dense seeds are desirable both for the farmer and the consumer, a 'win–win' situation.

Genetics

The micronutrient traits are reasonably stable across environments and the genetic control is relatively simple. Traits controlled by single major genes, often dominant, are common for both absorption of individual micronutrients and for their transport within the plant (Graham 1984, Graham et al 1999, 2000). For example, high zinc, manganese and copper density in cereals is controlled by two, one and one genes, respectively, and selection is easy in early generations using the ICP spectrometer (see Graham et al 1997, 1999). In the case of zinc and for all traits in polyploid species, multi-genic control is the rule. In diploids, about four loci control agronomic zinc efficiency, and we have identified three loci for loading of zinc into barley grains (P. Lonergan, unpublished results). Thus, the genetics are somewhat more complex than previously described but as the total analysis of elements in grains integrates these largely additive traits, progress seems to be quite feasible. The fact that there is a strong correlation among different elements in grain suggests that a common transport system is important, making progress easier. More than 20 molecular markers have already been described for micronutrient efficiency and loading traits.

Bioavailability

Initially, simple tests of bioavailability in rats have been done on the best material from the germplasm screening (Welch et al 1999). The criteria for these feeding trials were that the grains should be intrinsically labelled with isotope supplied to the plants throughout grain development so that all storage forms of minerals in grain are equally labelled, and secondly that the animals tested should be mildly deficient in the target nutrient. The latter requirement ensured that inducible systems present in the gut for more efficient absorption of the limiting nutrient under deficiency were operating during the test. In replete animals and humans, the high efficiency, inducible absorptive mechanisms are shut down to save energy and to avoid toxicity. The absorption efficiency (%) for iron and zinc from bean and rice lines so far tested has been fairly constant so that the amount of iron absorbed was linearly related to the total iron present, and similarly for zinc. This means that the simple ICP spectrometric measurement of total nutrient in grain is a good indicator of iron and zinc absorbed by rats.

Rats are not considered an ideal model for the human gut but parallel studies of absorption by human colon cell lines (Caco-2 cells, Glahn et al 1996) have given similar results to the rats. These results are encouraging, but tests in humans are essential to prove the validity of the breeding strategy. Initially, a small linear study of a village family in the Philippines was carried out. Six members were monitored for dietary intake over a period of 2 months with local rice, IR64, as the staple. After a blood sample was taken to establish the baseline iron status, the high iron, high zinc rice selected in the breeding program at the International Rice Research Institute was introduced. After another 2 months of similar diet, blood tests were again taken and significantly higher iron status was seen for several indices in the mother and children (but not the father whose iron status was always adequate). A similar study of young women in a convent ($n = 27$) followed the family study with similar results (L. del Mundo, personal communication). Another more extensive study bridging several convents has been planned to include more extensive control groups than previously. While the jury is still out on the proof of benefit, the results to date are encouraging.

Conclusion

The extent of micronutrient deficiencies in the human population, the risk of permanent brain damage from them and the speed of increase of the problem are evidence that current medical-based strategies are not working for billions of people out of contact with clinics. The truly sustainable strategy is based on agricultural systems delivering the required nutrients though the food supply. Proof of concept is now nearly complete. Our pre-breeding studies reviewed briefly here and more extensively elsewhere (Welch & Graham 1999, Graham et al 2001) strongly suggest that an alliance of agricultural and health scientists will be needed along with part of the resources now going to less sustainable strategies. This alliance will be challenged to deliver within a new paradigm for agriculture and its promise of improving the nutritional value of staple foods to have impact on all sectors of society, rich and poor, young and old.

References

Allaway WH 1986 Soil–plant–animal and human interrelationships in trace element nutrition. In: Mertz W (ed) Trace elements in human and animal nutrition. Academic Press, Orlando, FL, p 465–488

Anonymous 1992 International conference on nutrition: world declaration and plan of action for nutrition. Food and Agricultural Organization of the United Nations/World Health Organization, Rome, p 1–42

Anonymous 1994 The challenge of dietary deficiencies of vitamins and minerals. In: Enriching lives: overcoming vitamin and mineral malnutrition in developing countries. World Bank, Washington, DC, p 6–13

Brunson AM, Quackenbush FW 1962 Breeding corn with high provitamin A in the grain. Crop Sci 2:344–347

Christian P, West KP Jr 1998 Interactions between zinc and vitamin A: an update. Amer J Clin Nutr (suppl) 68:435S–441S

Darnton-Hill I, Cavelli-Sforza LT, Volmanen PVE 1992 Clinical nutrition in East Asia and the Pacific. Asia Pacific J Clin Nutr 1:27–36

García-Casal MN, Layrisse M, Solano L et al 1998 Vitamin A and β-carotene can improve nonheme iron absorption from rice, wheat and corn by humans. J Nutr 128:646–650

Georgievskii VI, Annenkov BN, Samokhin VT 1982 Mineral nutrition of animals. Butterworths, London

Glahn RP, Wien EM, Van Campen DR, Miller DD 1996 Caco-2 cell iron uptake from meat and casein digests parallels in vivo studies: Use of a novel in vitro method for rapid estimation of iron bioavailability. J Nutr 126:332–339

Graham RD 1984 Breeding for nutritional characteristics in cereals. Adv Plant Nutr 1:57–102

Graham RD, Ascher JS, Hynes SC 1992 Selecting zinc-efficient cereal genotypes for soils of low zinc status. Plant Soil 146:241–250

Graham RD, Senadhira D, Ortiz-Monasterio I 1997 A strategy for breeding staple-food crops with high micronutrient density. Soil Sci Plant Nutr 43:1153–1157

Graham RD, Senadhira D, Beebe SE, Iglesias C, Ortiz-Monasterio I 1999 Breeding for micronutrient density in edible portions of staple food crops: conventional approaches. Field Crop Res 60:57–80

Graham RD, Welch RM, Bouis HE 2001 Addressing micronutrient malnutrition through enhancing the nutritional quality of staple foods: principles, perspectives and knowledge gaps. Adv Agron 70:77–142

Gregorio GB, Senadhira D, Htut T, Khush GS, Graham RD 2000 Genetic variability for iron content in rice. Poster paper P-132, Abstracts of the International Rice Genetics Symposium, IRRI, Los Banos, October 22–27 2000. International Rice Research Institute, Manila, Philippines, p 294

Hodges RE, Saurberlich HE, Canham JE et al 1978 Hematopoietic studies in vitamin A deficiency. Am J Clin Nutr 31:876–885

House WA, Welch RM 1989 Bioavailability of and interactions between zinc and selenium in rats fed wheat grain intrinsically labeled with ^{65}Zn and ^{75}Se. J Nutr 119:916–921

Iglesias C, Mayer J, Chavez L, Calle F 1997 Genetic potential and stability of carotene content in cassava roots. Euphytica 94:367–373

Khachik F, Bernstein P, Garland DL 1997 Identification of lutein and zeaxanthin oxidation products in human and monkey retinas. Invest Opthamol Vis Sci 38:1802–1811

Moussavi-Nik M, Rengel Z, Hollamby GJ, Ascher JS 1997 Seed manganese (Mn) content is more important than Mn fertilisation for wheat growth under Mn-deficient conditions. In: Ando T, Fujita K, Mae T, Matsumoto H, Mori S, Sekiya J (eds) Plant nutrition for sustainable food production and environment. Kluwer Academic, Dordrecht, p 267–268

Muhilal, Parmaesih D, Idjradinata YR, Muherdiyantiningsih, Karyadi D 1988 Vitamin A fortified monosodium glutamate and health, growth and survival of children: a controlled field trial. Am J Clin Nutr 48:1271–1276

Panth M, Shatrugna V, Yashodara P, Sivakumar B 1990 Effect of vitamin A supplementation on hemoglobin and vitamin A levels during pregnancy. Br J Nutr 64:351–358

Prasad AS 1984 Discovery and importance of zinc in human nutrition. Fred Proc 43:2829–2834

Rengel Z, Graham RD 1995 Importance of seed Zn content for wheat growth on Zn-deficient soil. I. Vegetative growth. Plant Soil 173:259–266

Rosser JM, Khachik F, Graham RD 1999 Poster paper, 12th International Carotenoid Symposium, Cairns 18–23 July 1999

Simon PW 1992 Genetic improvement of vegetable carotene content. In: Bills DD, Kung SY (eds) Biotechnology and nutrition. Butterworth-Heinemann, Boston, p 291–300

Solomons NW, Russell RM 1980 The interaction of vitamin A and zinc: implications for human nutrition. Amer J Clin Nutr 33:2031–2040

Udomkesmalee E, Dhanamitta S, Sirisinha S et al 1992 Effect of vitamin A and zinc supplementation on the nutriture of children in Northeast Thailand. Amer J Clin Nutr 56:50–57

Weir RG, Hudson A 1966 Molybdenum deficiency in maize in relation to seed reserves. Aust J Exp Agric Anim Hus 6:35–41

Welch RM 1986 Effects of nutrient deficiencies on seed production and quality. Adv Plant Nutr 2:205–247

Whittaker P 1998 Iron and zinc interactions in humans. Am J Clin Nutr 68:442S–446S

Welch RM, Graham RD 1999 A new paradigm for world agriculture: meeting human needs; productive, sustainable, nutritious. Special volume. Welch RM, Graham RD (eds). Field Crops Res 60:1–10

Welch RM, House WA, Beebe S, Senadhira D, Gregorio G, Cheng Z 2000 Testing iron and zinc bioavailability in genetically enriched bean (*Phaseolus vulgaris* L.) and rice (*Oryza sativa* L.) using a rat model. Food Nutr Bull, in press

Yilmaz A, Ekiz H, Gultekin I, Torun B, Karanlik S, Cakmak I 1997 Effect of seed zinc content on grain yield and zinc concentration of wheat grown in zinc-deficient calcareous soils. In: Ando T, Fujita K, Mae T, Matsumoto H, Mori S, Sekiya J (eds) Plant nutrition for sustainable food production and environment. Kluwer Academic, Dordrecht, p 283–284

DISCUSSION

Bouis: I have few points to amplify what Robin Graham has been talking about. First, this whole area of micronutrient malnutrition is quite new. In the last 15 years there has been a sea-change in the attitudes of nutritionists towards nutritional problems in developing countries. This started with vitamin A trials that were undertaken in Indonesia. They gave a high dose of vitamin A to 10 000 pre-school children, and a placebo to another 10 000 children. This reduced the child mortality rate by 30%. They did eight other careful studies in other countries and found consistent results. This gives an idea of the importance of these micronutrients in the diet. One reason why plant breeding is such a powerful way of tackling this issue is cost. It costs about US$2.50 per year for a pregnant woman to take iron supplements. The World Bank estimates that this is a good investment, and that there will be a 20-fold return, because of the amelioration of the debilitating effects of iron deficiency. But if there are 20 million anaemic pregnant women in India each year, it will cost US$50 million, which is hard to find each year. This is where the plant breeding comes in: $50 million is more than enough to breed plants with enhanced trace minerals. It is much more cost effective to address these problems through a plant breeding strategy.

Mazur: Is anything known about the genetics of these loci?

Graham: Yes, there are a lot of major genes identified in micronutrient studies in plants. The more we look into it, however, the more we find there is a multitude of minor genes that will enhance the trait. I should qualify this: I am talking about the genetics of the control of absorption of nutrients from the soil where the efficiency traits are independently inherited for each element. The genetics of transport of these nutrients to the grain seems to involve new traits, which we are only just beginning to map. In barley we found three loci which enhance the transport and loading of zinc into the grain (P.F. Lonergan, personal communication), and there are four quantitative trait loci (QTLs) associated with loading of iron into rice grains (G. Gregorio, personal communication).

Mazur: So the mapping work is in barley?

Graham: Yes, but it is also taking place at IRRI on rice.

Mazur: Is the plan to develop markers for breeding, or will there be work to clone the genes as well?

Graham: In barley, our first objective is to develop markers, and we then intend to check them out in wheat.

Khush: The work on rice is for breeding.

G.-L. Wang: What is the genetic variation among the rice germplasm? You showed that in aromatic varieties there is higher iron content than in conventional rice.

Graham: That is true. Of all the sources of high iron in grain that we have found in rice, a large percentage of these are aromatic types.

Khush: This association is fortuitous, however. There are aromatic rices without high iron, and there are some high iron rices that are not aromatic.

Beyer: Are these varieties also low in phytic acid? This is a major problem in the bioavailability: high iron content alone may not do the job.

Graham: That is a hot potato. Generally, if they are high in iron they tend to be slightly higher in phytic acid. We are not sure whether we are dealing with multiple aleurone layer types here. In maize, multiple aleurone layer types are enhanced in concentrations of all the minerals and also phytic acid. The arguments for phytic acid are exceedingly complex. It is now emerging as an important signal transduction factor and is already well accepted as important in suppressing colon cancer. It has now been found in every living cell. It is found in high concentrations in all seeds, and we are not convinced that we should alter the phytic acid level, although many nutritionists would like to eliminate it. We have done just one study with low phytate maize, and grown these seeds on low phosphate soil. They are much less vigorous, which is what we would have predicted, even though the total phosphorus concentration in the low phytate types is the same. It is an availability issue. At the moment we are adopting a strategy of not going down the low phytate route, while maintaining a watching brief on the value of low-phytate mutants.

Leach: It seems that increasing the ability of the plants to take up micronutrients is an achievable target. But soon you would reach the limit of the fertility of the soil: no matter how well these plants can accumulate ions, if there is nothing in the soil they can't take it up. At what level does this become a problem?

Graham: With these heavy metals, it is the chemistry in the soil. The solubility product of various hydrated iron oxides is somewhere between 10^{-35} to 10^{-55}. The planet is replete with iron, it is just a matter of bioavailability and the ability of the plant to do the chemistry at the root–soil interface to solubilize the iron or other transition metals. All soils are in micronutrient balance from inputs of dust, rain, agrochemicals and even the wear and tear on farm implements; it is an issue of availability. If these nutrients were not as insoluble as they are, they would all be in the oceans by now.

Gale: One of the reasons that the high β-carotene golden rice has so much attention is that it did appear to be a solution that was not attainable by conventional breeding. Are there any other targets among the micronutrients that might be targets for transgenic strategies?

Graham: I would like to make the general point that human nutrition is a very difficult discipline, because people are not good subjects for experimental studies. Setting goals is difficult. There are all sorts of targets appearing in the literature, such as folate and vitamin E. Shintani & DellaPenna (1998) were able to shift the isomers in oil seeds from γ-tocopherol to α-tocopherol, which is more effective as vitamin E, without changing the total tocopherol in the oilseed. This is an example of a 'soft' molecular intervention that is particularly appealing. With folate, the latest news is an indication that women who are taking high doses of folate during pregnancy are now showing higher levels of twinning. This would be a highly undesirable outcome. Again, the human nutrition data are not giving us clear signals on where we should go. I think there is tremendous potential for biotechnology to tweak our food system, but getting the proper data for setting our strategy is the key.

Okita: What is the distribution of micronutrients between the aleurone, pericarp and bulky endosperm in rice?

Graham: A lot of the phytate and the heavy metals are found in the aleurone, which is milled away. This is an important issue for our project. The early data suggest that even though we might mill away more than half of the iron from the brown rice, the concentration advantage of the milled high-iron rice over the standard rice may be still twice as much. There are also interactions between the genotype and the milling process. In some varietal comparisons, high-iron rice might be milled and lose a much higher percentage than standard rice, and vice versa. This is one more parameter that we have to account for.

Ku: Robin Graham, you didn't say much about selenium, which is essential for human and animal health. What do we know about the selenium content in rice? I saw startling statistics from a survey among 60 countries: uptake of selenium is inversely proportional to the cancer rate.

Graham: The US Plant, Soil and Nutrition Laboratory (Ithaca, NY) published a map of selenium in the soils in the USA. An oncologist obtained this map and found that it resembled the map of the distribution of cancer in the USA. This is what precipitated the selenium theory of the aetiology of cancer. Since then extensive studies have taken place that have generated some impressive correlations between selenium and cancer incidence, so these data are becoming quite solid. However, in the Arizona study, suppression of cancer was achieved by supplying four times the published recommended dietary allowance of selenium. There are two levels of interest in selenium. The first is in the absolute deficiency, such as we see in China and Zaire and many other countries, and second there are the levels that seem to be required for cancer suppression. There are few data on selenium levels in rice, but we are scaling up our project to generate these data and to look for genetic variability. For every trait that we have looked for so far in germplasm banks we have found useful variability, so I am sure that we will find this for selenium also. I think we can address the question of selenium through varietal selection. This will be advantageous in that all these interactions and synergies can be exploited. Selenium is essential for the deposition of iodine in the thyroid hormone molecule. Seleno-proteins are intimately involved in thyroid regulation. If we can deal with these things through the food system, then I feel that this will be the most successful approach.

Khush: There are reports that selenium content is very high in certain irrigation waters in parts of Bangladesh and Pakistan.

Gale: Robin Graham, I noticed that in your data you showed an organic/conventional farming comparison. In the UK, commercial organic yields are generally less than half of conventional farming yields. Were the reduced yields taken account of in your data? Do you think there is a place for commercial organic farming in rice production in Southeast Asia?

Graham: I don't think I can answer that for rice. The data I gave were concentrations, so they are yield independent. We eat on a concentration basis: if iron is low, you don't eat twice as much food. My general feeling about organic systems is that their lower productivity is a concern while the population pressure is on the land. The point has been made that if we didn't have these highly productive cereal systems then vast acreages of forests would have to be cleared to meet current food production needs, and this would be an environmental disaster. In line with the ideas of L. T. Evans (1998), I feel that organic farming systems are an unattainable ideal for a global population in excess of about three billion.

Khush: Organic rice farming would be a recipe for disaster.

References

Evans LT 1998 Feeding the ten billion: Plants and population growth. Cambridge University Press, Cambridge

Shintani D, DellaPenna D 1998 Elevating the vitamin E content of plants through metabolic engineering. Science 11:2098–2100

Biosynthesis of β-carotene (provitamin A) in rice endosperm achieved by genetic engineering

Salim Al-Babili, Xudong Ye*, Paola Lucca*, Ingo Potrykus* and Peter Beyer[1]

*University of Freiburg, Center for Applied Biosciences, D-79104 Freiburg, Germany, and *Institute for Plant Sciences, Swiss Federal Institute of Technology, CH-8092 Zurich, Switzerland*

Abstract. To obtain a functioning provitamin A (β-carotene) biosynthetic pathway in rice endosperm, we introduced in a single, combined transformation effort the cDNAs coding for (1) phytoene synthase (*psy*) and (2) lycopene β-cyclase (*β-lcy*; both from *Narcissus pseudonarcissus* and both under control of the endosperm-specific glutelin promoter), with (3) a bacterial phytoene desaturase (*crt*I, from *Erwinia uredovora* under constitutive 35S promoter control). This combination covers the requirements for β-carotene synthesis, and yellow, β-carotene-bearing rice endosperm was obtained in the T_0 generation. However, further experiments revealed that the presence of *β-lcy* was not necessary, since *psy* and *crt*I alone were able to drive β-carotene synthesis as well as the formation of further downstream xanthophylls. This finding could be explained if these downstream enzymes are either constitutively expressed in rice endosperm or are induced by the transformation, e.g. by products derived therefrom. Based on results in *N. pseudonarcissus* as a model system, a likely hypothesis can be developed that *trans* lycopene or a *trans* lycopene derivative acts as an inductor in a kind of feedback mechanism stimulating endogenous carotenogenic genes.

2001 Rice biotechnology: improving yield, stress tolerance and grain quality. Wiley, Chichester (Novartis Foundation Symposium 236) p 219–232

Rice is the major staple food for millions of people. It is generally consumed in its milled form with outer layers (pericarp, tegmen, aleurone layer) removed. The main reason for milling is to remove the oil-rich aleurone layer, which turns rancid upon storage, especially in tropical and subtropical areas. As a result, the edible part of rice grains consists of the endosperm, filled with starch granules and protein bodies, but it lacks several essential nutrients for the maintenance of

---

[1]This chapter was presented at the symposium by Peter Beyer, to whom correspondence should be addressed.

health, such as carotenoids exhibiting provitamin A activity. Thus, rice contributes to vitamin A deficiency, a serious public health problem in at least 26 countries including highly populated areas of Asia, Africa and Latin America.

The mildest form of vitamin A deficiency is night blindness; however, continuous deficiency causes the symptoms of xerophthalmia and keratomalacia, leading to irreversible blindness. In Southeast Asia it is estimated that 5 million children develop xerophthalmia every year, of which quarter of a million eventually go blind (Sommer 1988). Furthermore, vitamin A deficiency is correlated to fatal afflictions such as diarrhoea, respiratory diseases and childhood diseases such as measles. Vitamin A deficiency negatively alters the incidence of such illnesses, their duration and severity (West et al 1989). According to statistics compiled by the UNICEF, the diets of an estimated 124 million children worldwide are deficient in vitamin A (Humphrey et al 1992). Improved nutrition could prevent 1–2 million deaths annually among children aged 1–4 years and additional 0.25–0.5 million deaths during later childhood (West et al 1989). Since attempts to overcome the problem by oral application of vitamin A are problematic (Pirie 1983, Sommer 1989), mainly due to the lack of infrastructure and educated medical staff, alternatives are urgently required. Such an alternative for highest-risk countries could be to supplement the major staple food, rice, with provitamin A. This can only be achieved by recombinant technologies rather than conventional breeding, due to the lack of rice cultivars producing this provitamin in the endosperm. Since on one hand, the transformation of rice is well established and since, on the other hand, the carotenoid biosynthetic pathway has been molecularly identified during recent years, it appeared feasible to introduce the complete provitamin A (β-carotene) biosynthetic pathway into rice endosperm by genetic engineering.

We have shown previously that immature rice endosperm synthesizes the early intermediate geranylgeranyl diphosphate. This compound is not solely devoted to carotenogenesis but can be used as a substrate to produce the uncoloured carotene phytoene by expressing the heterologous enzyme phytoene synthase in rice endosperm (Burkhardt et al 1997). We report here on the successful introduction of the complete provitamin A biosynthetic pathway into rice endosperm by genetic engineering, yielding yellow-coloured rice grains to be used to defeat illnesses caused by provitamin A deficiency.

Results and discussion

To engineer the pathway towards β-carotene formation the complementation of this transgenic plant with three plant enzymes is necessary, namely phytoene desaturase and ζ-carotene desaturase catalysing the introduction of two double bonds each, and lycopene β-cyclase. Alternatively, to reduce the transformation

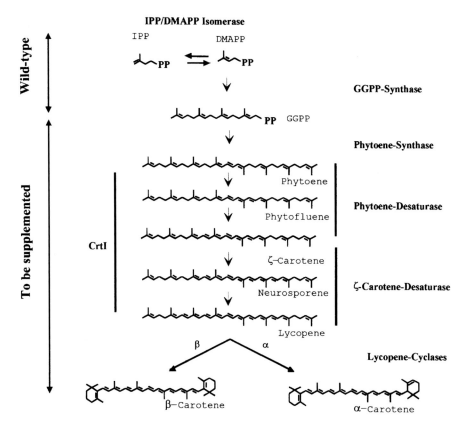

FIG. 1. Provitamin A biosynthetic pathway. The names of enzymes are given. CrtI denotes a bacterial carotene desaturase capable in performing all necessary desaturation reactions for which two enzymes are required in plants. Arrows indicate the prenyllipid biosynthetic capacity of wild-type rice endosperm and the necessary reaction sequence to be completed to yield provitamin A.

effort by reducing the number of enzymes required, certain bacterial carotene desaturases can be used which are capable of introducing all four double bonds required (see Fig. 1).

Initially, we introduced all genes into immature rice embryos (TP 309) singly by particle bombardment, aiming at unifying all transgenes into a single plant by subsequent crossing. However, this approach was not successful, mainly due to the deleterious integration pattern frequently produced by this transformation technique, as revealed by Southern hybridization analysis. Therefore, *Agrobacterium*-mediated transformation of precultured rice immature embryos was used, in order to install the entire β-carotene biosynthetic pathway into rice endosperm in a single transformation effort. Three vectors, schematically

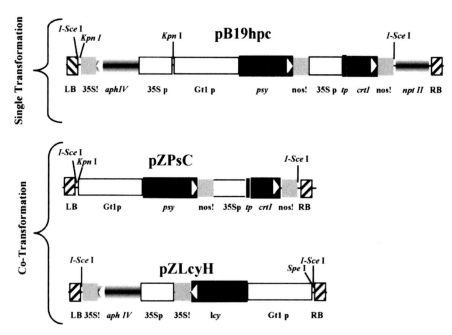

FIG. 2.   DNA constructs used in single transformations and co-transformations. RB, LB, right and left borders; tp, transit peptide from pea ribulose bisphosphate carboxylase; !, terminator; p, promoter; gt, glutelin. DNA sequences coding for carotenoid biosynthetic enzymes are given in black; *psy*, phytoene synthase; *crt*I, bacterial carotene desaturase; *lcy*, lycopene β-cyclase.

depicted in Fig. 2, were constructed. pB19hpc combining the sequences for a plant phytoene synthase (*psy*) originating from daffodil (*Narcissus pseudonarcissus*; Acc. No. X78814; Schledz et al 1996) with the sequence coding for a bacterial phytoene desaturase (*crt*I) originating from *Erwinia uredovora* (Acc. No D90087), the two being placed under the control of the endosperm-specific glutelin (Gt1) and the constitutive CaMV 35S promoter, respectively. The phytoene synthase cDNA contained a 5′-sequence coding for a functional transit peptide, as was demonstrated earlier (Bonk et al 1997), while the *crt*I gene was fused to the transit peptide sequence of the pea Rubisco small subunit (*tp*), as constructed by Misawa et al (1993). This plasmid should thus direct the formation of lycopene in the endosperm plastids, the site of geranylgeranyl diphosphate (GGPP) formation.

To complete the β-carotene biosynthetic pathway, co-transformations were carried out employing two vectors, one (pZPsC) carrying *psy* and *crt*I, like in pB19hpc but lacking the selectable marker *aph*IV expression cassette, the other (pZCycH) providing under glutelin promoter control the sequence coding for the enzyme lycopene β-cyclase, originating from *N. pseudonarcissus* (Acc.

No. X98796, Al-Babili et al 1996). As with phytoene synthase, lycopene β-cyclase carried a functional transit peptide allowing plastid-import (Bonk et al 1997). The combination of both plasmids should thus direct β-carotene formation in rice endosperm.

A total of 800 precultured rice immature embryos were inoculated with *Agrobacterium* LBA 4404 /pB19hpc. Fifty hygromycin-resistant plants were then analysed by Southern blot (not shown) using DIG-labelled internal fragments from *psy* and *crt*I DNA as probes. Meganuclease I–*Sce*I digest released a *c.* 10 kb insertion comprising the *aph*IV, *psy* and *crt*I expression cassettes. *Kpn*I was used to estimate the insertion copy number. All tested lines carried the transgenes and most of the plants revealed single insertions, but in some cases multiple insertions were observed.

For co-transformation about 500 precultured immature embryos were inoculated with an *Agrobacterium* mixture of LBA4404/pZPsC carrying *psy* and *crt*I genes and LBA4404/pZCycH containing *lcy* and *aph*IV as the selectable marker. Co-transformed plants were identified by Southern hybridization. To probe for the presence pZPsC, the same restriction was applied as above using the respective probes, whereas the presence of the pZCycH expression cassettes was screened separately by probing I-*Sce*I and *Spe*I-digested genomic DNA with internal *lcy* fragments. All 60 randomly selected regenerated lines were positive for *lcy,* among which 12 plants were co-transformed with pZPsC as shown by the presence of the expected fragments; 6.6 kb for the I-*Sce*I-excised *psy* and *crt*I expression cassettes from pZPsC and 9.5 kb for the *lcy* and *aph*IV genes from pZCycH. Like the transformation from above, 1–3 transgene copies were predominant in co-transformed plants. Ten plants harbouring all four introduced genes were transferred into the greenhouse for setting seeds. All plants from all transformations described here showed a normal phenotype as well as normal fertility.

Mature F_0 seeds from transformed lines and from control plants were air dried, dehusked and, in order to isolate the endosperm, treated with emery paper for 8 h on a shaker. In most cases the transformed endosperms exhibited clearly notable yellow colour, indicating carotenoid formation. The pB19hpc single transformants showed a clear 3:1 (coloured/non-coloured) segregation pattern, whereas the pZPsC/pZCycH double transformants (Fig. 3B) showed a wider deviation in segregation, as expected. To our surprise, the pB19hpc single transformants, although equipped for lycopene (red) synthesis, were not distinguished in colour compared to the pZPsC/pZCycH double transformants equipped for β-carotene (yellow) synthesis.

Seeds from individual lines (1 g, each) were ground to a fine powder and extracted to complete decolourization with acetone. The combined extracts were quantified photometrically and analysed qualitatively by HPLC (Fig. 3). The

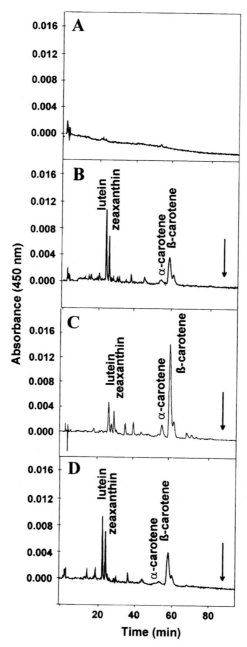

FIG. 3.   HPLC analysis of wild-type and transformed rice endosperm on a C_{30} reversed phase column. (A) untransformed control which is carotenoid-free. (B) single transformant, (C, D) co-transformants. The arrow indicates the position of lycopene in the system used.

carotenoid pattern observed with the pB19hpc single transformants explained the phenotype that we noted visually. None of these lines accumulated detectable amounts of lycopene. Instead, the pathway was completed to form β-carotene and even lutein and zeaxanthin were formed to some extent, resulting in a carotenoid pattern that is qualitatively quite similar to the one present in green leaves. This suggests that the lycopene $\alpha(\varepsilon)$- and β-cyclases as well as the hydroxylase are either constitutively expressed in rice endosperm or that the expression of these downstream enzymes is induced by lycopene formation or by products derived therefrom (see below).

The pZPsC/pZCycH double transformants exhibited a more variable carotenoid pattern (Fig. 3). This ranged from phenotypes that are similar to the ones from the single transformations to others that contain β-carotene as almost the only carotenoid. Our line z11b is such an example (see Fig. 3B) also representing up to now the 'winner' in quantitative terms. A carotenoid content of 1.6 μg/g endosperm was determined. Considering the fact that the F_0-generation of seeds now analysed is not homozygous with respect to the transgenes (so that a segregation between colourless, intermediate-coloured and strongly-coloured grains takes place), it can be assumed that this line as well as several others will safely reach our anticipated goal of 2 μmole/g endosperm (Burkhardt et al 1997). From a nutritional point of view it is not yet clear whether lines producing provitamin A (β-carotene) or lines possessing additionally zeaxanthin and lutein are to be selected, since it turns out during recent years that these xanthophylls are present in the eye's macula and their deficiency may contribute to macular degeneration, leading to blindness (Landrum et al 1997). In this respect lutein and zeaxanthin may therefore represent valuable compounds exhibiting vitamin character. Work has been initiated to cover these aspects as well as safety considerations that will begin as soon as there is sufficient homogeneous material available.

Preliminary evidence has accumulated to provide an explanation for the unexpected carotenoid pattern in the transgenic rice seeds. Currently it cannot be excluded that the transformation using the bacterial *crt*I-gene promotes a hitherto unknown feedback mechanism enabling the transcriptional activation of carotenogenic genes. The effector may be the lycopene itself (or products derived therefrom) that is all-*trans* configured when being formed by the bacterial enzyme, while the two plant desaturases yield a poly-*cis* configured lycopene, termed prolycopene (Bartley et al 1999). In a working hypothesis, prolycopene represents a biosynthetic intermediate, while *trans* lycopene may act as an initiator of this feedback mechanism. To test this, we took advantage of the chemical compound CPTA (2-[chlorophenylthio]triethylamine hydrochloride) that acts as a lycopene cyclase inhibitor and leads to the accumulation of *trans* lycopene (Beyer et al 1991). CPTA-treatment thus mimics with respect to *trans*

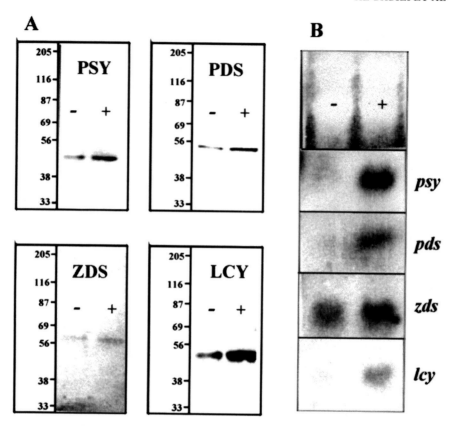

FIG. 4. Northern and western blot analysis of CPTA-treated and untreated daffodil flowers. (A) Western Blot. Total protein (10 μg) was isolated from treated (+) and untreated (−) flowers and after SDS-PAGE transferred to nitrocellulose membranes. The blots were then developed using homologous antibodies directed against the carotenogenic enzymes as given. (B) Northern blot. Total RNA (20 μg) was isolated from CPTA-treated (+) and untreated (−) flowers. The immobilized RNA was repeatedly stripped to allow subsequent hybridization with different probes. PSY, phytoene synthase; PDS, phytoene desaturase; ZDS, ζ-carotene desaturase; LCY, lycopene cyclase.

lycopene formation our rice single transformation using plasmid pB19hpc. CPTA was administered to daffodil flowers, which then turn reddish within 8 h due to lycopene accumulation. However, concomitantly the carotenoid content was increased two-to-three times over the untreated controls. Northern blots conducted with probes directed against four carotenogenic mRNAs as well as Western blots conducted with the respective homologous antibodies showed that both the specific mRNAs examined as well as the specific protein amounts were markedly increased over the untreated controls (Al-Babili et al 1999; Fig. 4).

This result cannot be explained by the well-known action of CPTA as a lycopene cyclase inhibitor but indicates the presence of a novel regulatory mechanism. Utilizing the transgenic rice in hand, further work is in progress to clarify molecularly as to whether this mechanism works here as well.

Conclusions

In a 'proof-of concept', we have shown that it is possible to establish a biosynthetic pathway *de novo* in rice endosperm enabling the accumulation of provitamin A. Many variations of the applied technology appear now feasible and work is in progress to optimize the yellow rice now in our hands. This relates to e.g. the use of different structural genes or the use of different selectable marker genes. With the observation that the lycopene β-cyclase is not necessary to achieve provitamin A synthesis it may even be possible to remove the selectable marker from co-transformants by breeding techniques (see Fig. 2). The observation that a regulatory pathway may be involved calls for in-depth biochemical and molecular biological analyses, which will be carried out as soon as sufficient material becomes available. Then, studies on the bioavailability of the provitamin, transfer of the trait into elite varieties, and risk assessments etc. will be carried out in collaboration with other research institutes.

Acknowledgements

This work was supported by the Rockefeller Foundation (I.P. and P.B.) and by the European Community Biotech Program (P.B., FAIR CT96, 1996-1999), the Swiss Federal Office for Education and Science (I.P.) and by the Swiss Federal Institute of Technology (I.P.)

References

Al-Babili S, Hobeika E, Beyer P 1996 A cDNA encoding lycopene cyclase (Accession No. X98796) from *Narcissus pseudonarcissus* L. (Plant Gene Register 96-107). Plant Physiol 112:1398

Al-Babili S, Hartung W, Kleinig H, Beyer P 1999 CPTA modulates levels of carotenogenic proteins and their mRNAs and affects carotenoid and ABA content as well as chromoplast structure in *Narcissus pseudonarcissus* flowers. Plant Biol 1:607–612

Bartley GE, Scolnik PA, Beyer P 1999 Two *Arabidopsis thaliana* carotene desaturases, phytoene desaturase and zeta-carotene desaturase, expressed in *E. coli*, catalyze a poly-*cis* pathway to yield prolycopene. Eur J Biochem 259: 396–403

Beyer P, Kröncke U, Nievelstein V 1991 On the mechanism of the lycopene isomerase/cyclase reaction in *Narcissus pseudonarcissus* L. chromoplasts. J Biol Chem 266:17072–17078

Bonk M, Hoffmann B, von Lintig J et al 1997 Chloroplast import of four carotenoid biosynthetic enzymes *in vitro* reveals differential fates prior to membrane binding and oligomeric assembly. Eur J Biochem 247:942–950

Burkhardt P, Beyer P, Wünn J et al 1997 Transgenic rice (*Oryza sativa*) endosperm expressing daffodil (*Narcissus pseudonarcissus*) phytoene synthase accumulates phytoene, a key intermediate of provitamin A biosynthesis. Plant J 11:1071–1078

Humphrey JH, West KP Jr, Sommer A 1992 Vitamin A deficiency and attributable mortality among under 5-year-olds. WHO Bull 70:225–232

Landrum JT, Bone RA, Joa H, Kilburn MD, Moore LL, Sprague KE 1997 A one year study of the macular pigment: the effect of 140 days of a lutein supplement. Exp Eye Res 65:57–62

Misawa N, Yamano S, Linden H et al 1993 Functional expression of the *Erwinia uredovora* carotenoid biosynthesis gene *crtl* in transgenic plants showing an increase of β-carotene biosynthesis activity and resistance to the bleaching herbicide norflurazon. Plant J 4:833–840

Pirie A 1983 Vitamin A deficiency and child blindness in the developing world. Proc Nutr Soc 42:53–64

Schledz M, Al-Babili S, von Lintig J, Rabbani S, Kleinig H, Beyer P 1996 Phytoene synthase from *Narcissus pseudonarcissus*: functional expression, galactolipid requirement, topological distribution in chromoplasts and induction during flowering. Plant J 10:781–792

Sommer A 1988 New imperatives for an old vitamin (A). J Nutr 119:96–100

Sommer A 1989 Large dose vitamin A to control vitamin A deficiency. In: Walter P, Brubacher G, Stähelin H (eds) Elevated dosages of vitamins: benefits and hazards. Hans Huber Publishers, Toronto, p 37–41

West KP Jr, Howard GR, Sommer A 1989 Vitamin A and infection: public health implications. Annu Rev Nutr 9:63–86

DISCUSSION

Parker: Do you know why your original crossing strategy didn't work?

Beyer: I have no idea. There have been problems with expression when using the ballistic approach mainly due to deleterious integration patterns. The change to *Agrobacterium*-mediated transformation was a good choice, because this gave us nice integration patterns and good expression.

Leach: Do you find an increase in carotenoids in green parts of the plants?

Beyer: We looked at that. We see no change in the green parts of our plants, although one of the transferred genes is under constitutive promoter control.

Okita: Have you ever sectioned one of the seeds to see whether there is uniform synthesis of β-carotene?

Beyer: It is hard to tell: I have the impression that it is quite uniform, although the GT1 promoter is more active in the outer layers of the endosperm.

Khush: Could you see any patches on the grain samples?

Beyer: No. The colour is not just restricted to some layers of the endosperm.

Presting: What is the nutritional significance? Is it enough vitamin A to make a difference? Are you planning to look at some of the interactions that Robin Graham described between zinc, iron and vitamin A?

Beyer: Keep in mind that provitamin A is not the vitamin. The vitamin the result of a dioxygenase-cleavage which occurs in animals. Vitamin A is then normally converted into an ester, which represents the effective compound. The amount required is measured in retinyl equivalents (RE). The dosage required to avoid the problems of provitamin A deficiency is $100\,\mu g$ RE per day. This would necessitate $2\,\mu g$ β-carotene/g uncooked rice. Our most successful rice lines contain the provitamin in that range. We get $1.4–1.8\,\mu g/g$, which is almost the $2\,\mu g$ that we wanted. But you have to consider that these are numbers from analyses carried out with the segregating lines so, according to Mendelian laws, we should exceed that value in the homozygous lines. However, bioavailability is still an issue that remains to be investigated as soon as there is sufficient material available.

Ku: What is the efficiency of co-transformation in the last experiment you did?

Beyer: It was surprisingly high. Among the resistant plants, they were 20% co-transformed.

Ku: Earlier, you said something about segregation: what was the problem there? In our case when we crossed two lines with the two genes, we only needed to look at about 50 seedlings to find one homozygous for both genes.

Beyer: We have the 3:1 segregation pattern in the single transformant. The co-transformants are simply unpredictable, as one may imagine. This awaits further analysis.

Leach: Is there any β-carotene in the plastids in the roots?

Beyer: No.

Gale: Peter Bayer, your work is tremendously important for public perception of genetic modifications. What is your deployment strategy for the golden rice?

Beyer: Everything is looking very encouraging at the moment. Most important is to keep the rice on the beneficial track on which it begun. This is not an easy task to do. We must, for instance care for the intellectual property rights (IPR) issue. We tried to settle this ourselves, but this has not been possible because of our lack of expertise in this field. Luckily we received help from the ISAAA, a humanitarian organization that is involved in the transfer of technology from developed countries to developing countries. They are carrying out an IPR audit as a first step, revealing that about 70 patents must be taken into consideration. Meanwhile we have received considerable assistance from other parties but no contracts have been signed as yet. Because of this I can't disclose much. However, I can read out a few statements which explain our joint position. 'Agreement has been reached between the inventors of golden rice and a number of private sector organizations. Contractual agreements will be completed shortly. Central to the agreement is that the investors are able to honour all their original concepts of making the technology free to resource-poor farmers in developing countries. This technology is a gift to these people from the inventors and the Rockefeller Foundation. The principal partner company will assist by providing specialist technical support to the investors and recipient country institutions. The Rockefeller Foundation which initiated this project almost 10 years ago sees this as a good example of public–private partnership in making available the technology to contribute to the elevation of vitamin A deficiency, induced blindness and malnutrition. Other not-for-profit organisations have been and will be invited to support the project too, to further enhance the efforts in this latest stage of co-operation.'

Mazur: This shows that freedom to operate can be a messy issue. How many different parties are represented in those 70 patents?

Beyer: There are several parties involved. But the number of parties is reducing with all the mergers that are taking place!

Leach: What is the acceptance going to be of rice that is yellow? I remember that the Rockefeller Foundation did a study many years ago where they tried to get people in Asia to eat yellow rice, and they refused because they associated yellow rice with tainted rice.

Beyer: There may be some lack of acceptance when people do not see a benefit. But this yellow rice will be available to people who urgently need improved nutrition. Certainly acceptance is also linked to the perception of potential risks. I would like to add that environmental and risk assessment studies will be carried out. Metabolic profiling studies will also be essential. Potentially, interfering with signal pathways could have unforeseen results, which may even have a positive impact, but we need to double check the content of the seeds.

Bouis: With regard to the dissemination strategy, I have spent a good part of the last five years trying to push this idea of breeding for nutrition. I think the problems can be overcome, but we have to be clear about the different steps once the breeding is done. The first step in the dissemination strategy is to incorporate the nutritional characteristics in the highest-yielding and most profitable varieties. The farmers will not adopt these varieties because they are nutritious, but because they are profitable. We have some evidence that the trace minerals are compatible with high yields. We have no reason to believe that the β-carotene is going to harm yields, but it will not promote yields either. The second step is ensuring consumer acceptance. There are many strategies that communications experts have for trying to get people to adopt things that they are not used to. This will be a cost that will have to be borne by governments, convincing consumers to eat yellow rice. With the trace minerals, we don't think it is going to change the consumer characteristics of the rice, so we are not as worried about consumer acceptance. The third and most difficult issue is that of bioavailability. We have to establish that these things are really going to help peoples' nutrition. On the trace minerals, people are consuming perhaps 50% of the RDAs, and we think we can raise this by another 25% with the parameters we currently have. The data on vitamin A rice from Bangladesh suggest that at the present levels of β-carotene, this would increase the vitamin A intakes by 25%. The final problem is getting the money together to try these strategies. We have been making these arguments about the trace minerals for the last five years and have only been able to raise US$1.5 million for this strategy. It has been difficult.

Graham: On this acceptance issue, in Australia we have a wheat variety registration process, which involves bakers and millers. They are very conservative. They want white flour and say that this is what the market demands. We also have a maverick breeder who came up with a rather creamy-floured wheat of a beautiful agronomic type. He found a market for yellow alkaline noodles in Malaysia, and got this wheat past the registration board in this way. We have found that the creamy-coloured bread made from this wheat has a sweeter smell than the white and appears to hold its freshness, so there is now a boutique industry developing around this variety. A more sophisticated market is evolving.

Khush: I should add that in the Indian subcontinent, in certain preparations saffron colour is added to rice. This natural yellow rice would have much higher acceptance there. There is one problem in consumer acceptance of golden rice: when rice is spoiled by poor storage it becomes yellowish. Consumers might therefore equate golden rice with spoiled rice. However, I feel that these problems can be addressed through public education.

Presting: You showed fairly high co-transformation efficiency in your system. I am interested in getting feedback from anyone who is doing transformations with the intent of public release on whether it would be a good strategy to use co-transformation and eventually cross-out the selectable marker. Is this something we should worry about when we make these transgenic organisms?

Beyer: This is certainly an issue. I am told by people more experienced than myself in transformation that in most cases the integration is in one locus, so we can't cross this out. It is probably a matter of how many lines we look at. It would be taking such an effort to result in selectable marker-free plants. In addition, we have a visible phenotype, which assists us greatly in doing these kind of things.

Presting: Which selectable marker did you use in your experiments?

Beyer: Hygromycin, but we are aiming to use the mannose-phosphate isomerase systems in new transformations.

Ku: I would add that the antibiotics we use for selection of transformed plants are not commonly used in human disease treatments. However, there is public concern about their use.

Elliott: I particularly appreciate the sensitivity of Maurice Ku's last comment. The anti-GM crops campaigns have raised public awareness of our use of selectable markers by suggesting that such markers do threaten the health of animal and human populations that eat GM crops. While we may believe that this is nonsense, we should not respond to accusations of sloppy molecular biology with reflex negatives. The commercial crops do not need the markers, so we should exploit transformation systems for generating marker-free transgenic plants (Joersbo & Okkels 1996, McCormac et al 1999, Yolder & Goldsbrough 1994). Our own pBECKs 2000 vectors have been designed to address the need for a 'clean gene' system which allows the redundant marker genes to be removed from the genome of the transgenic plant after selection (McCormac et al 1999).

References

Joersbo M, Okkels FT 1996 A novel principle for selection of transgenic plant cells: positive selection. Plant Cell Reports 16:219–221

McCormac AC, Elliott MC, Chen DF 1999 pBECKS 2000: a novel plasmid series for the facile creation of complex binary vectors, which incorporates 'clean-gene' facilities. Mol Gen Genet 261:226–235

Yoder JI, Goldsbrough AP 1994 Transformation systems for generating marker-free transgenic plants. BioTechnology 12:263–267

Developing transgenic grains with improved oils, proteins and carbohydrates

Barbara J. Mazur

DuPont Agricultural Enterprise, PO Box 80402, Wilmington, DE 19880-0402, USA

Abstract. DuPont has developed cereals and oilseeds with improved proteins, carbohydrates, and oils for food, feed, and industrial applications. Products which have been or will be introduced include corn and soybeans with increased oil content, improved oil composition, increased amino acid content, altered protein content and functional qualities, altered starch composition, reduced oligosaccharide content, increased sucrose content, and combinations of these traits. These products have been developed using both mutation breeding and molecular biology-based transgenic approaches. We have also worked on improving the underlying technologies in order to accelerate product introductions. Gene discovery has been expedited through a genomics program that now has a database of more than two million sequences from a variety of plants, insects and microbes. Plant cell transformation for elite lines of crop species is being addressed through production laboratories with high throughput processes and through technology improvements. High-throughput, rapid and small-scale assays for biochemical parameters are used to identify plants carrying traits of interest. Small-scale functionality analyses, in which grains are broken down into their component parts and assayed for functional properties, indicate which seeds carry a trait of commercial value. Finally, a number of DNA marker systems are being used to accelerate trait introgression timelines.

2001 Rice biotechnology: improving yield, stress tolerance and grain quality. Wiley, Chichester (Novartis Foundation Symposium 236) p 233–241

DuPont's biotechnology programs have focused on improving grain quality and crop protection traits in corn, soybeans, wheat and rice, through both breeding and molecular technologies (Mazur 1995). The products from this work include seeds with improved agronomic traits, and grains with improved quality traits for food ingredient and animal feed products. In order to commercialize these products, we have acquired or allied with a number of businesses in the food distribution system, including breeding, processing and distribution companies.

The first step in creating transgenic crops for commercial users is to identify traits that could contribute added value for growers, processors or consumers.

Trait identification is a difficult process that must balance end-user evaluations of the potential value of a trait with the research and development costs required to create the novel seed. Adding chemically or biologically synthesized ingredients to grain samples to create prototype products for testing can expedite this evaluation. Alternatively, a mutant line with the desired properties can be identified for product testing. For example, the value of the thermal and oxidative stability of a high oleic acid oil from a mutant soybean line was judged to be sufficient to justify a transgenic approach (Kinney & Knowlton 1998). The transgenic plant had significant advantages over the mutant prototype in that it possessed agronomic hardiness and environmental stability in addition to a high oleic phenotype.

Product development timelines for novel grain products are long, and shortening these timelines is essential to create a viable business opportunity. As much as 8–12 years have been needed to develop novel grain products. We have spent considerable effort to condense these timelines, through the use of genomic technologies for gene discovery, high throughput transformation for transgenic plant development, and molecular breeding and fast cycle breeding for variety and inbred development (Mazur & Tingey 1995).

Formerly, gene isolation for transgenic plant development was performed using a combination of biochemical and genetic technologies to clone single candidate genes. In recent years this work has been accelerated through the application of a genomics approach, in which cDNA libraries are created from expressed genes from a variety of plants and other organisms, and clones are sequenced in parallel using high throughput automated processes and entered into a common database (Rafalski et al 1998). The DuPont EST database includes genes from corn, soybeans, wheat, rice, insects, and microbes, as well as a number of biologically diverse species that contribute genes with unique specificities.

The addition of libraries from biodiverse sources to the database has enabled the identification of genes specifying enzymes with catalytic properties that are not present in crop species. Examples of this approach include the cloning of diverged oleic acid desaturase genes from *Vernonia*, *Momordica* and *Impatiens* (Hitz 1998, Cahoon et al 1999). By randomly sequencing 3000 expressed sequence tags (ESTs) from cDNA libraries from developing seeds of each of these species, genes could be identified that encoded enzymes that catalyse the formation of epoxide or conjugated fatty acids from linoleic acid. Transgenic expression of full-length copies of these genes in a soybean embryo model system resulted in the accumulation of the fatty acids. This work embodies a new paradigm for gene discovery, in which enzymes with useful specificities are identified through a survey of biodiverse organisms, followed by a random sequencing genomics program and identification of the gene of interest.

This approach relied on a homology comparison to identify the targeted genes. Bioinformatics tools are a critical component of a genomics strategy, both to assign gene functions through homology searches, and to analyse and visualize the vast amounts of data generated during expression profiling experiments. Gene and protein profiling experiments survey gene expression under particular developmental, spatial, growth or environmental conditions, for the purpose of uncovering the function of unknown genes (Mazur 1999). Several gene expression profiling technologies are being used to narrow candidate gene searches from tens of thousands of genes to hundreds of genes. In one example, a selected set of 1500 ESTs was spotted onto glass slides, and the DNA arrays were hybridized with fluorescent dye-labelled cDNA probes synthesized from kernel and embryo mRNA isolated at successive stages of embryo development (J. A. Rafalski, personal communication). Patterns of coordinated gene expression in metabolic pathways could be discerned, and the data were used to select candidate genes for work to alter seed oil content. Protein expression profiling, with protein samples from developing seeds arrayed on two dimensional gels and individual proteins identified by mass spectrometry, is adding a further dimension to these analyses (Yates 1998).

While these experiments were performed on a pre-selected set of genes, other technologies are being used to simultaneously query expression of all genes in a cell. These technologies include a bead-based technology for massively parallel signature sequencing developed by Lynx Therapeutics, in which a 17–24 base signature sequence can be read from an array of micro beads, each with 10^5 attached copies of a single cDNA (Brenner et al 2000). The number of times that a particular sequence is sampled in a population of up to half a million beads indicates the relative expression level of that gene under the assayed conditions. The advantage of this technology is that there is no pre-selection of the genes to be assayed, all genes can be assayed simultaneously, and sequences are read directly, making secondary gene identification protocols unnecessary.

Genetic crosses play a vital role in assigning gene function and identifying candidate genes of interest. Creating linked genetic and physical maps, with cDNAs localized at regular intervals along the map, allows mutant genes and quantitative trait loci to be isolated by a map-based cloning strategy. Importantly, such strategies allow the genes responsible for a particular trait to be directly identified based on their map position, without the biases associated with assigning genes to roles in a pathway that is inherent in molecular biology-based approaches. Further, the genome syntenies identified between cereals can be used to clone genes in one cereal based on anchored maps in a related grass (Gale & Devos 1998), with the caveat that microsynteny between species is not always maintained, as shown for the rice and maize Adh genes by Tarchini et al (2000).

The employment of plant populations carrying genetic tags that cause the inactivation or ectopic activation of genes augments these strategies. An example comes from work to create high lysine grain for use in food and feed products.

Expression of a deregulated gene for dihydrodipicolinic acid synthase in maize led to a 25-fold increase in embryo free lysine (Falco et al 1995). However, analysis indicated that free lysine was being lost through catabolism, with the accumulation of unwanted catabolites and with yield declines. A Mutator element-tagged maize population (Briggs & Meeley 1999) was screened to identify lines with Mutator insertions in the lysine ketoreductase gene, which was believed to encode the enzyme responsible for the lysine catabolism. When identified, such lines had a 10-fold increase in free lysine levels, and when crossed with the lysine overproducing lines, a 50-fold increase in free lysine was achieved versus wild-type maize, with lysine accumulation in embryos as well as in endosperm, and a reduction in lysine catabolites (S. C. Falco, personal communication). This work validated the lysine ketoreductase gene as a target for transgenic cosuppression. The lines could be used directly, or single locus, dominant high lysine transgenes could be created which combine seed-specific dihydrodipicolinic acid synthase expression and seed-specific lysine ketoreductase suppression.

The use of a combination of these technologies to identify a gene for isoflavone synthase provides another example of how a database search was narrowed from hundreds of thousands of ESTs to a few candidate genes (Jung et al 2000). Elevated isoflavone content in soybeans is a nutritional trait of value, as isoflavone intake has been implicated in the reduction of cholesterol levels and in reduced incidences of certain cancers, osteoporosis and heart disease. The gene encoding isoflavone synthase, which catalyses the first committed step for the pathway and was believed to be a cytochrome P450, had proven difficult to identify; all other enzymatic steps in the predicted pathway could be assigned ESTs from our soybean database. Bioinformatic tools were used to identify all P450 ESTs in the database, and to screen them according to their tissue and environmentally induced patterns of expression. Several genes that were expressed in fungally challenged plants were shown to synthesize the isoflavone genistein in a yeast complementation assay. This activity was then confirmed by expression of the transgene in a soybean embryo model system, and by stable transformation into soybean and *Arabidopsis*.

These and other screening technologies could be augmented by the development of truly high throughput plant transformation systems, which would allow direct analysis of a larger number of candidate genes than is presently practical. In the absence of such throughput, we have relied on model systems, such as the soybean embryo model system and BMS maize cell lines, for more rapid predictions of transgene activity. We have also invested considerable effort in streamlining our transformation processes. Evaluation of optimal transformation and growth conditions, coupled with the construction of dedicated production facilities, has now enabled a high level of transgene throughput.

The above examples lead to the identification of transgenic product candidates based on biochemical parameters. However, product functionality testing, to evaluate the characteristics of a transgenic grain in a particular product application, is an essential confirmatory step in the development process (Mazur et al 1999). The requirements for functionality testing vary widely. For soybean ingredient products, the seeds must first be broken down into meal, flour, concentrate, isolate and/or oil fractions, in laboratory-scale versions of equipment that mimic commercial-scale processes. An example of such product testing comes from the development of soybeans with reduced oligosaccharide content. Biochemical analyses showed that these lines had reduced stachyose content and elevated sucrose content. In order to ascertain their utility for soy beverages, they were evaluated with consumer testing panels, where they were shown to have digestibility profiles equal to that of cow's milk, as measured by hydrogen evolution in the subjects. Lines with altered protein composition due to cosuppression of the gene for the seed storage protein conglycinin have functionality characteristics similar to egg albumin, when assayed by tests for fat and water binding and emulsification and rheological properties (Fader & Kinney 1997). Lines with high oleic acid content were shown to have extended shelf life and high temperature frying attributes when tested in a tiered system of assays, starting with simple laboratory-based active oxygen measurements and proceeding to large-scale customer-based frying tests (Kinney & Knowlton 1998). High oleic acid oil was also demonstrated to have utility as a hydraulic fluid after supplementation with additives, by measuring hydraulic pump metal wear in a low volume fluid power test (Glancey 1998). Finally, for feed ingredient valuation, grain must be fed to poultry and livestock in a battery of trials, to determine the added nutritional value of the grain for each particular species at different stages in the growth cycle.

Once a trait of value is identified, it then must be moved into elite, high yielding lines for commercial production, because any performance losses due to crop yield reductions will reduce the profitability of the new trait. Further, depending upon commercialization plans, lines for production in specific maturity zones must be employed. Having access to the highest yielding inbred lines and varieties is essential; such lines are typically the product of decades of plant breeding. Because baseline crop yields continue to improve annually, moving the trait into the most competitive seed lines is a continuous challenge. Using DNA markers at early stages of backcrossing for trait introgression can accelerate the breeding process by one to two years. DNA markers can also be used to improve crop yields, by retrospectively analysing pedigrees to identify loci that contribute to yield in the top varieties and inbred lines.

For successful commercialization, it is important to combine multiple added value traits in a single grain product. Because differentiated grains require separate collection, storage, shipping and handling, added costs are incurred in maintaining the necessary identity preservation channels. Combining several traits in a single grain product spreads the handling costs among multiple products, and makes speciality grains a commercial possibility. However, the combination of traits must reach similar markets or require similar acreage to add commercial value; a high acreage oil trait must be combined with a high acreage meal trait to be beneficial commercially. An example is the high oil corn trait, which has sufficient value to enable other traits of lesser value, such as high lysine, to be successfully commercialized with it.

Finally, a number of other factors play a role in a successful commercialization process. Intellectual property protection must be applied for early in the discovery process. Conversely, freedom-to-operate with respect to competitive patents must be negotiated. Regulatory approvals are increasingly lengthy, with varying requirements from one country to the next; the processes to obtain multiple national approvals must be initiated early in the development process. Contract acreage for crop production must be sought, and quality control of all aspects of production and processing must be high. Commercial partnerships must be negotiated throughout the product development chain, and must include trait discovery and development, crop production, shipping and handling, processing, product manufacturing, and marketing components. This has increasingly led to vertical business integration throughout the value chain, with technology driving the evolution of business systems as companies have sought to introduce products. Finally, public trust and acceptance of biotechnology-based plants must be gained, and the consumer educational process must be global in nature for these products to achieve success.

From this work, foods with desirable health attributes are being developed. Examples include oils with increased monounsaturated fatty acids and decreased saturated fatty acids that have oxidative and thermal stability without hydrogenation and without concomitantly introduced *trans* fatty acids. Flours isolated from high oleic soybeans have improved stability. Flours from low stachyose and high sucrose soybeans have improved digestibility, due to replacement of anti-nutritional oligosaccharides with sucrose. Flours with altered ratios of the major seed storage proteins have differentiated functional properties, including water and oil binding, viscosity, flavour and texture, which make them suitable for particular food ingredient applications. Grains with increased concentrations of essential amino acids such as lysine and methionine (Falco et al 1995) have value for foods and feeds. Soybeans with increased isoflavone content can have health benefits, whereas those with reduced isoflavones can have taste and organoleptic attributes. Feeds with environmental benefits are also being developed. An example of this is the introduction of corn and soybeans with high available phosphorus and low phytate content; such grains have a reduced environmental impact following animal feeding due to the reduced bound phosphorus content of the animal waste. And starches with altered branching patterns and altered amylose and amylopectin ratios, and consequently altered rheological properties, have applicability in a variety of food and industrial products (Hubbard 1997). The majority of these products could not have been created without the assistance of biotechnology tools, and offer new health benefits for consumers, food production benefits for a growing population, and environmental benefits for the ecosystem.

References

Brenner S, Williams SR, Veermas EH et al 2000 *In vitro* cloning of complex mixtures of DNA on microbeads: physical separation of differentially expressed cDNAs. Proc Natl Acad Sci USA 97:1665–1670

Briggs SP, Meeley RB 1999 US Patent 5962764 Functional characterization of genes

Cahoon EB, Carlson TJ, Ripp KG et al 1999 Biosynthetic origin of conjugated double bonds: production of fatty acid components of high-value drying oils in transgenic soybean embryos. Proc Natl Acad Sci USA 96:12935–12940

Fader GM, Kinney AJ 1997 Suppression of specific classes of soybean seed protein genes. Patent No. WO97/47731. Issued 18 December 1997

Falco SC, Guida T, Locke M et al 1995 Transgenic canola and soybean seeds with increased lysine. Biotechnology (NY) 13:577–582

Gale MD, Devos KM 1998 Comparative genetics in the grasses. Proc Natl Acad Sci USA 95:1971–1974

Glancey JL 1998 Development of a high oleic soybean oil-based hydraulic fluid. SAE Technical Paper 981999. Society of Automotive Engineers, Warrendale, PA

Hitz WD 1998 Fatty acid modifying enzymes from developing seeds of *Vernonia galamenensis* and fatty acid desaturase gene DNA sequences. US Patent 5846784. Issued 8 December 1998

Hubbard NL, Klein T, Broglie K 1997 Novel starches via modification of expression of starch biosynthetic enzyme genes. Patent No. WO97/22703. Issued 26 June 1997

Jung W, Yu O, Lau S-MC et al 2000 Identification and expression of isoflavone synthase, the key enzyme for biosynthesis of isoflavones in legumes. Nat Biotechnol 18:208–212 (erratum: 2000 Nat Biotechnol 18:559)

Kinney AJ, Knowlton S 1998 Designer oils: the high oleic soybean. In: Harlander S, Roler S (eds) Genetic engineering for food industry: a strategy for food quality improvement. Blackie Academic, London, p 193–213

Mazur BJ 1995 Commercializing the products of plant biotechnology. Trends Biotechnol 13:319–323

Mazur B 1999 Technology issues in plant development. Nat Biotechnol 17:BV9–BV10

Mazur BJ, Tingey SV 1995 Genetic mapping and introgression of genes of agronomic importance. Curr Opin Biotechnol 6:175–182

Mazur B, Krebbers E, Tingey S 1999 Gene discovery and product development for quality traits. Science 285:372–375

Rafalski JA, Hanafey M, Miao G-H et al 1998 New experimental and computational approaches to the analysis of gene expression. Acta Biochim Pol 45:929–934

Tarchini R, Biddle P, Wineland R, Tingey S, Rafalski A 2000 The complete sequence of 340 kb of DNA around the rice Adh1-Adh2 region reveals interrupted colinearity with maize chromosome 4. Plant Cell 12:381–391

Yates JR 1998 Database searching using mass spectrometry data. Electrophoresis 19:893–900

DISCUSSION

Parker: This might be a naïve question. Companies have to make profits. However, there is a massive need in the developing world for products that are going to come on to the market. Are people in companies, as well as in the public sector, thinking about strategies to make these products available at an affordable price without diminishing the profits of the companies? Is this something that companies are thinking about?

Mazur: It is a complex issue. You can segment products into different markets. For example, poultry feed in the USA could be segmented away from feeding operations in another part of the world. It is possible to create differential markets. Once we make a decision on whether we are going to be in the rice business, we will be able to determine what we do with traits that we have developed for corn and soybean. We might make these traits available to the public sector, but we may want to retain traits for particular market segments.

Nevill: I agree. It is clear that there will be areas where there is commercial interest, and other areas where things go more into the public domain because we don't see that we are going to have a profitable situation. I think it is important to discuss these issues on a project-by-project level and not globally, since it is a question of the end-use of a particular technology, rather than its overall development. It is therefore possible for various partners to gain a benefit from what we are doing. Rice is probably a good example. Both DuPont and Novartis have been debating about whether there really is a business position here. Clearly, if there is, it won't be such a broad interest as might be expected in other crops. In other words, if we went and did something it would be more in high-value areas, under which circumstances there could easily be a spin-off for IRRI or others working in the public domain.

Mazur: One example that has come to our attention is the high lysine trait in rice: this might be a trait that we would make available.

G.-L. Wang: It seems that most of your projects are focused on developing value-added traits for corn and soybean. In the public sector, because we cannot compete with companies, do you think we should focus more on input traits, such as yield and disease resistance?

Mazur: I focused on the value-added grain traits, as this is an area that we at DuPont have worked on historically. We worked on grain traits because this market was identified by our business group, and because we didn't have a seed company at the time. Pioneer Hi-Bred has a seed business and has focused more heavily on agronomic traits and crop protection traits. Now that we have merged there has been a complementarity in these programmes.

Okita: What kind of reception does the poultry and cattle industry have for GM products?

Mazur: It has not been an issue in the past. When we have talked to the animal integration companies they have been anxious to have products produced through this technology. It offers them the promise of lower cost.

Salmeron: Could you comment on the challenges you have found in setting up identity preservation for these different products? Do you see differences in the different crop market segments?

Mazur: Identity preservation systems have been implemented for both corn and soybean. It has been a big challenge, but it is getting easier. We have set up near-infrared instruments at grain elevators to measure quality traits so that farmers can be paid on a quality basis. Implementation has been through commercial partnerships with grain traders. Increasingly this will become the way of the future.

Summing-up: cutting-edge science for rice improvement — breakthroughs and beneficiaries

John Bennett

Plant Breeding, Genetics and Biochemistry Division, International Rice Research Institute, DAPO Box 7777, Metro Manila, Philippines

Looking back over the papers presented in this symposium, I am struck that the mandate of IRRI, to help poor rice farmers and consumers in low-income countries, will be directly facilitated by the science on display. The reason is simple: breakthroughs in plant genetics lead to improved seed, and improved seed is one of the best ways of delivering the benefits of scientific research to poor farmers and consumers (Fig. 1).

Feeding the poor: the role of modern genetics

In his opening address, economist Mahabub Hossain of IRRI highlighted both the achievements of rice research over the last 40 years and the challenges for the future. The high-yielding semi-dwarf rice varieties of the Green Revolution allowed global production to grow at a faster annual rate than population between 1970 and 1985. As a result, mass starvation was avoided and the poor were able to afford more rice than previously. Between 1985 and 1998, however, rice production grew at a slower rate than population, creating a new urgency for further productivity gains. Over the next 20 years, production will have to grow by 40% to keep the world's rice bowl full. On the debate as to whether the world has a food distribution problem or a food production problem, Hossain observed that low-income countries do not have the financial resources to buy rice and other staples from exporting countries. There is also a limited rice production capacity outside Asia. Countries in the region must themselves strive to increase rice production, especially higher-quality rice for which there is a rising demand. Although irrigated rice will continue to account for more than 70% of total rice production in Asia, the vast majority of poor farmers grow rice under rainfed lowland and upland conditions, with all the unfavorable features that characterize those environments. Research must focus on both the irrigated and the rainfed ecosystems.

Breeder Gurdev Khush of IRRI, chairman of the symposium, noted that rice provided 23% of the calorie intake of the world's human population. In Asian countries the percentage was even higher (29–73%). In the future, more rice would have to be produced from less land and that would require scientific discoveries permitting higher productivity, including more resistance to biotic and abiotic stresses.

Hossain and Khush proposed four strategies for ensuring food security and alleviating poverty over the next 40 years. These strategies were as follows:

1. Raise yield potential in favorable environments.
2. Reduce yield gaps in unfavorable environments.
3. Raise income of small farmers through enhanced grain quality.
4. Alleviate malnutrition through increased micronutrient content.

The pursuit of these strategies will involve many scientific disciplines but plant breeding will continue to play the central role. Khush described the use of conventional breeding to break the current yield record of ~ 10 tonnes ha^{-1} in a season under tropical conditions, held by inbred indica cultivars. This record would be exceeded by inbred tropical japonica cultivars currently under development in IRRI's New Plant Type project. The heterosis of indica/tropical japonica hybrids would then boost yields higher still. Further gains could also be expected from several current applications of biotechnology, including attempts to manipulate photosynthesis and starch metabolism.

Rice breeders now have many new molecular tools to accelerate their work or increase its scope, in addition to providing a deeper understanding of rice biology. As the symposium would illustrate, some of these molecular insights derive from studies on rice itself, while other breakthroughs originate with other cereals such as maize and even from dicotyledonous plants such as *Arabidopsis* (Fig. 1).

Genomics: a global view of genetic change and constancy

Until about 10 years ago, most plant research was conducted on individual genes, proteins or metabolic pathways or on 'black boxes' consisting of tissues or whole plants. Although highly informative, this reductionist approach was recognized by its practitioners as being somewhat limited. Results from different laboratories were difficult to integrate into a single holistic vision. Then, in 1983, the advent of genetic engineering allowed the outputs of biochemical and molecular analysis to be studied in a whole-plant context, bringing to an end the decades-long contention between reductionism and holism. At the same time, DNA-based

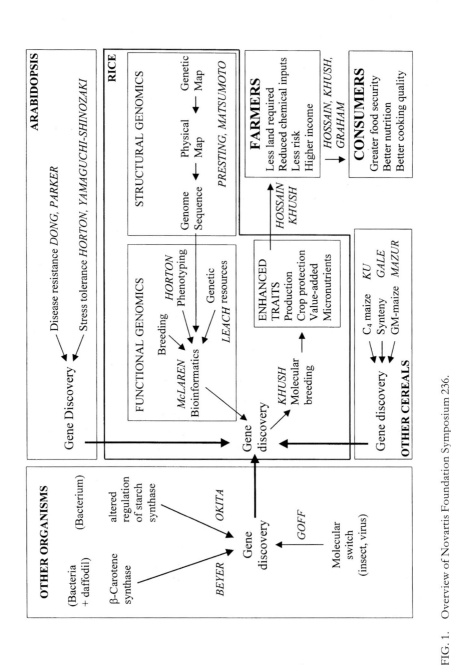

FIG. 1.   Overview of Novartis Foundation Symposium 236.

genetic maps of plants ushered in a new era of genomic research, in which genes controlling a wide range of traits could be viewed simultaneously in the context of their locations on the chromosomes of plants. The production of various types of large-insert DNA libraries enabled scientists to convert genetic maps into physical maps and opened the way to the complete sequencing of plant genomes. Genomics and proteomics now give scientists the analytical tools they need to look at gene expression and the regulation of metabolism in an integrated fashion.

The small genome size of rice gives this species the status of a model cereal for genetic and physical mapping. Gernot Presting (Clemson University, USA) and Takashi Matsumoto (National Institute of Agrobiologocal Resources, Japan) described progress by the International Rice Genome Sequencing Project to prepare a physical map of rice cultivar Nipponbare using large-insert YAC, BAC and PAC libraries. About 11% of the genome has been sequenced, including 5% at Clemson through BAC end sequencing and 5% at NIAR through cDNA sequencing. The results already in hand show clearly many of the dynamic rearrangements and duplications that have occurred in the rice genome, including the flux of retrotransposons and other mobile elements. An important aspect of publishing the DNA sequence is providing annotation. The goal of annotation is to identify the tens of thousands of segments of the sequence that corresponding to genes, mobile elements and other functional regions, and to use the tools and resources of bioinformatics to explore their functions. Gene discovery is one of the key outputs of the structural and functional phases of rice genomics (Fig. 1).

For Mike Gale (John Innes Centre, UK), it is the constancy of the cereal genome that is remarkable, not just its fluidity. His message was simple: rice breeders, wheat breeders and maize breeders should now consider themselves cereal breeders. In the 60 million years since the divergence of modern cereals from their last common ancestor, the genomes of these key crops have diverged in haploid chromosome number (between 5 and 12) and in haploid DNA content (from 430 Mbp to more than 2000 Mbp). And yet the order of the $\sim$ 30 000 cereal genes has remained remarkably constant (synteny). Local departures from synteny notwithstanding, the chromosomal locations of many genes in one cereal are now predictable from their locations in other cereals, greatly simplifying the task of gene discovery for cereal breeding (Fig. 1). The news about synteny between cereals and *Arabidopsis* was less encouraging. Gale reported that three laboratories had found that synteny across the monocot–dicot divide was minimal and perhaps limited to just a few genes at a stretch.

The amount of genetic information in the rice genome can perhaps be appreciated, if we imagine printing out the complete 430 Mbp of the Nipponbare DNA sequence, when it becomes available in a few years time. Printed in the four-letter alphabet of DNA (A, C, G and T), the code would occupy the equivalent of

18 volumes of the 1004-page Manila telephone directory, if the tiny font size of the directory were retained. The task of handling this amount of sequence information and linking it with other types of information was the theme of IRRI's biometrician, Graham McLaren. By itself, the DNA sequence of the Nipponbare genome would be very unwieldy and comparatively uninformative. Its value would be enhanced enormously by a series of linkages to other databases and other types of information. The process of annotation of the sequence uses databases from other organisms to suggest putative functions for many of the rice genes. Data on the map location of genes conditioning specific traits allows the tools of functional genomics to be applied to identifying candidate genes for each trait. Bioinformatics also seeks to incorporate breeding data, phenotypic records and passport data of the $> 80\,000$ accessions in the IRRI Genebank.

Increasing yield: manipulating source and sink

There is probably no aspect of plant biology that is better understood at the molecular level than photosynthesis, and yet it is difficult to point to any increase in crop yield that has arisen from this understanding. The fact is that photosynthesis is merely the source of the carbohydrate required for yield in the grain. We need to learn much more about the sink and the transport and plant growth processes that intervene. The source–sink balance was a major theme of the symposium.

C3 plants such as rice experience saturation of photosynthesis at a much lower light intensity than C4 plants such as maize. They also waste more of their resources on photorespiration and transpiration. Could the yield potential of rice be increased through the gain of some features of C4 photosynthesis? Maurice Ku (Washington State University, USA) reported on the effects of transforming rice with maize genes encoding the C4 enzymes phospho*enol*pyruvate carboxylase (PEPC), pyruvate P_i dikinase (PPDK) and NADP-malic enzyme (NADP-ME). Rice plants overexpressing C4 PEPC exhibited a higher photosynthetic rate, a larger stomatal aperture and greater activities of carbonic anhydrase and Rubisco than control plants. The larger stomatal aperture may be due to the effects of carbonic anhydrase and PEPC on the potassium malate content of guard cells. Rice plants overexpressing C4-PPDK showed a similar range of properties. Overexpression of NADP-ME in rice increased tillering and stress tolerance. Plants containing two copies of all three genes had the highest photosynthetic rate, largest stomatal conductivity and highest internal CO_2 concentration. These plants showed a 12% increase in yield in small-scale tests. Nevertheless, C4 photosynthesis is most beneficial when accompanied by changes in leaf structure (Kranz anatomy) and compartmentation of C4 enzymes. Ku believes that it might be possible to identify a master switch in maize that induces all the anatomical and biochemical changes required for C4 syndrome.

Tom Okita (Washington State University, USA) argued that rice yield is limited by the sink capacity of the grain rather than by the source capacity of the leaves. He reported a 10% increase in seed weight as a result of transforming rice with a triple mutant of the bacterial *glgC* gene, encoding ADP-glucose pyrophosphorylase. The mutant AGPase was insensitive to allosteric inhibition by phosphate and 3-phosphoglycerate.

Peter Horton (Sheffield University, UK) focused on the effects of excess light intensity on rice leaves. Rice cultivars with erect leaves intercept light efficiently in the morning and afternoon but inefficiently at noon. This is seen as a mechanism for reducing light harvesting under high irradiance. Rice also dissipates excess excitation energy by non-photochemical quenching mechanisms such as those mediated by protonation of the light-harvesting apparatus and the xanthophyll cycle. In Horton's view, higher yields in dry and wet seasons may be obtainable by developing distinct cultivars for each season rather than expecting a single genotype to acclimate adequately in both seasons.

Fighting hidden hunger: human micronutrients in the rice grain

Carbohydrate and protein deficiencies are the most obvious causes of hunger, but micronutrient deficiencies (particularly of vitamin A, iron, zinc, iodine and selenium) cause 'hidden hunger' in more than half of the world's population. Micronutrient deficiencies are especially common among women and children. They cause preventable blindness, mental retardation, anemia and poor growth in children. Because of their ease of storage and high daily consumption, cereal grains offer an excellent delivery mechanism for the missing micronutrients.

Focusing on hidden hunger, Robin Graham (University of Adelaide, Australia) described the beneficial effects of abolishing iodine deficiency in a remote region of China through addition of iodide to the water used to grow wheat. He also drew attention to the high iron content of the milled grain of some traditional and improved rice cultivars. Preliminary results from human feeding trials indicate that much of the iron of these grains is available for absorption. Studies at IRRI showed that the high iron content of the grain is compatible with high yield.

Peter Beyer (University of Freiburg, Germany) described the science behind the production of rice with enhanced pro-vitamin A content in the endosperm. Two genes from daffodil (encoding the carotogenic enzymes phytoene synthase and lycopene cyclase) and one gene from the bacterium *Erwinia* (encoding phytoene desaturase) were inserted into rice cultivar Taipei 309. When expressed together, these genes led to the production of yellow grains containing carotenoids, predominantly pro-vitamin A. To make pro-vitamin A-enriched rice available to the millions of Asian children suffering from preventable blindness, three steps remain. Firstly, the three genes will be transferred to rice cultivars popular in Asia. Secondly, issues arising from intellectual property rights on transformation protocols and materials will be resolved. Thirdly, public unease about genetic engineering of food crops must be addressed.

Linking structural and functional genomics: inspiration from *Arabidopsis*

The spin-off to rice of research on *Arabidopsis* was well illustrated by several speakers. Peter Horton described a screening procedure that detected insertion mutants of *Arabidopsis* seedlings that were altered in photosynthesis. The screen was based on detecting chlorophyll fluorescence from leaves. The mutants identified so far include several affected in their ability to adapt to excess light.

Jane Parker (John Innes Centre, Norwich, UK) dissected some of the defence pathways and networks that protect *Arabidopsis* from viral, bacterial and fungal pathogens. Different pathogens can activate similar pathways, including a lipid signalling pathway that resembles the prostoglandin pathway in animals. Some mutants in this pathway induce a lesion mimic response. Another pathway is similar to the programmed cell death response of animals.

Mutants in the systemic acquired response pathway of *Arabidopsis* were studied by Xinnian Dong (Duke University, Durham, USA). These mutants include *cpr1-8* (in which pathogenesis related proteins are expressed constitutively) and *npr1* (in which PR proteins are not expressed). NPR1 is a transcription factor and transformation of rice with NPR1 of *Arabidopsis* conferred resistance to bacterial blight.

Jan Leach (Kansas State University, USA) described some of the molecular events underlying the interaction between rice and the bacterial blight pathogen, *Xanthomonas oryze*. Avirulence proteins of the pathogen enter host cells and interact with receptor proteins that elicit expression of defence response (*DR*) genes. Recombination among variant copies of the avirulence genes inside the bacterium creates new protein forms that escape detection by the host, leading to virulence. Leach and colleagues at IRRI and elsewhere are accumulating resources to develop an efficient screen to detect mutants with deletions in *DR* genes. The resources include a collection of ~ 30 000 deletion mutants in cultivar IR64. They also feature microarrays of candidate *DR* genes derived from metabolic and signal transduction pathways and from subtractive cDNAs libraries synthesized from mRNA of rice undergoing resistant and susceptible responses.

The theme of abiotic stress tolerance was developed by Kazuko Yamaguchi-Shinozaki (Japan International Research Center for Agricultural Sciences, Japan). She described the enhancement of drought, salinity and cold tolerance in *Arabidopsis* through transformation with a modified form of the gene encoding transcription factor DREB1A. This gene is normally activated by cold, whereas the gene encoding a related factor, DREB1B, is activated by salinity and drought. Both transcription factors bind to the dehydration-responsive element (DRE) of many genes, including *rd29A*. When *Arabidopsis* was transformed with DREB1A under the control of the CaMV 35S promoter, drought, salinity and cold tolerances were all enhanced but plant growth was stunted. When the same gene was introduced under the control of the *rd29A* promoter, multiple stress tolerances were observed without stunting. Rice homologues of these genes are now being isolated (Fig. 1).

From lab to field: lessons from the private sector

In several respects, IRRI and the other CGIAR centers have more in common with the private sector than with universities. Their goal is impact in farmers' fields, not simply proof-of-concept. It was therefore instructive to learn from two private sector managers about the impact of commercial and regulatory decisions on their biotechnology activities.

The focus of the paper by Steven Goff (Novartis Agricultural Discovery Institute, USA) was the development of molecular switches that place transgene expression in plants under the control of exogenous small molecules, such as the insect molting hormone, ecdysone. *Arabidopsis* was transformed with two genes, the first encoding the ecdysone receptor fused with the VP16 transactivation domain of the herpes simplex virus. The second gene encoded firefly luciferase and was under the control of the *cis*-element recognized by the ecdysone receptor. The luciferase gene was essentially inactive in the absence of inducer and was unresponsive to endogenous chemicals and environmental influences. However, spraying as little as 0.1 μM ecdysone gave a 50 000-fold induction of luciferase. The fact that ecdysone has already been cleared for agricultural spraying is an important feature because it removes a multimillion dollar approval hurdle.

Barbara Mazur (DuPont Agricultural Products Experimental Station, USA) used high lysine content and altered fatty acid composition of seeds as illustrations of how novel output traits are engineered into soybean and maize. Besides needing to make crucial molecular discoveries and inventions to arrive at working systems, DuPont had to devote considerable resources to ensure freedom-to-operate with genes, promoters and protocols. Another problem arose from the fact that the company's breeders are continually producing improved new varieties. It is therefore essential that DuPont's genetic engineering team is able to quickly introduce value-added traits into the new cultivars before they are superseded.

TABLE 1 Strategies, breakthroughs and beneficiaries

Strategies	Challenges	Breakthroughs	Constraints	Beneficiaries
Raise yield potential in favourable environments	Increasing population size, less land for rice production, need to reduce pollution	New plant type. Indica/japonica hybrids. C4 photosynthesis. Tolerance of excess light. Improved starch deposition.	Increase percentage grain filling. Keep source/sink in balance. Master control for C4 syndrome needed.	Greater food security for consumers. Higher incomes for farmers. Less environmental degradation.
Reduce yield gaps in unfavourable environments	Diseases, insects, drought salinity, excess light, submergence	Defence genes and pathways identified by functional genomics. Master control genes available.	Networks of genes and pathways, diversity of pests and diseases, environmental effects, evolutionary legacy.	Reduced chemical costs for farmers. Improved health and less economic risk for farmers. Cleaner food for consumers.
Raise income of small farmers through enhanced grain quality	Quality: palatability, aroma, milling and cooking quality	Functional genomics	Genetic complexity and subjectivity of traits	Higher incomes for rice farmers. Higher quality grain for consumers.
Alleviate malnutrition through increased micronutrient and amino acid content.	Fe, Z, I, Se, vitamin A, lysine	Transgenic rice with high vitamin A content Transgenic maize with high lysine. Mapping genes for high Fe content of grain. I from fertilizer enters grain.	Human nutrition research needed throughout project	Higher incomes for rice farmers. Improved nutrition for rice consumers.

Conclusions

The symposium began with a list of four strategies by which rice research could help low-income countries feed themselves and progress along the path of sustainable development. The strategies focused attention on crop productivity, crop protection, value-added traits and human nutrition. By the end of the symposium, the speakers had established beyond question that modern biology has the power to contribute significantly to each of these strategies (Table 1). Most progress has been made in relation to enhancing yield potential, reducing yield gaps and improving the micronutrient content of the grain, but there are still constrains that must be addressed. The progress in relation to grain quality has not been so impressive, but the advent of functional genomics gives scientists powerful new tools to examine these complex and subjective traits.

It is clear that public understanding of modern biology has not kept pace with these developments, leaving the people of developed and developing countries alike unable to assess the issues for themselves. Most people in developed countries have sufficient purchasing power to buy their food in packaged and processed forms from supermarkets. They are accustomed to seasonal fruits and vegetables being available all year round. They do not value the connection between modern agricultural research and the quantity, quality, availability and cost of their food. For most people in developing countries, the situation could not be more different. They must grow some or all of their own food or buy it from local farmers. They are very conscious that in the dry season or the wet season many foods are unavailable. In such circumstances, the achievement of a balanced diet can be difficult, especially for the poor. They are in a better position to appreciate the value of many of the new traits that modern biology can introduce into rice and other crops to benefit them as farmers or consumers.

The large agricultural biotechnology companies plan to launch a public awareness campaign on food safety. While this initiative is welcome, there can be no substitute for the direct demonstration to farmers and consumers of the benefits that they can expect from the products of molecular breeding within a regulatory system administered by competent officials. IRRI and the other CGIAR centers have an important role to play in developing case studies in biosafety and food safety of relevance to developing countries.

Index of contributors

Non-participating co-authors are indicated by asterisks. Entries in bold indicate papers; other entries refer to discussion contributions.

252

Subject index

A

AA genome species 98
abiotic stresses 12, 248
abscisic acid (ABA) 149, 177
acclimation ceiling 132
acclimation manipulation 117–134
ACeDB system 78
ADP glucose pyrophosphorylase *see* AGPase
Aegilops umbellulata 50, 55
AGPase 43, 142–144, 147, 148, 247
  mutant enzymes 143
Agrobacterium 104, 221, 223, 228
Amplified Fragment Length Polymorphisms
  (AFLP) 71
amyloplast genome 149–150
anchoring chromosome 10 contigs 21
anchoring contigs in silico 21–22
annotated sequence 38
annotations 34, 39
APETALA2 181
Arabidopsis 21, 34, 38, 50, 52, 58, 86, 96, 121,
  125, 130, 133, 140–142, 148, 151,
  155–158, 166, 176–189, 193, 196, 200,
  236, 243, 248–249, 345
Arabidopsis centromere 25
Arabidopsis mutants
  aba (ABA-deficient) 179
  abi (ABA-insensitive) 179
Arabidopsis thaliana 28–29, 72, 125, 154,
  165–175
ATP synthase 123
auxin 149

B

BAC 24–26, 27, 41, 46, 51, 202
  end-sequence analysis 14–15
  end-sequencing 14–15, 64
  fingerprinting 15
  libraries 14
bacterial artificial chromosome *see* BAC
Bambusoideae 50
banding pattern 16

barley 55, 137
Berkley *Drosophila* Genome Project (BDGP)
  29–30
Big Dye Primer 34
Big Dye Terminator 31, 34
bioavailability of micronutrients 211–212
biochemical analyses 237
biochemical pathways 67
biochemistry 72–74
bioinformatics 59–84, 246
bioinformatics tools 234–235
biological data 67
biological databases 66–67
Biomedical Information Science and
  Technology Initiative (BISTI) 66
biotechnology tools 239
BLASTN analysis 21–22
blindness, preventable 248
branching enzymes 148
Brassica 58
brown plant hopper 204

C

C3 pathway 101
C3 photosynthesis 97, 99, 102, 115, 246
C4 photosynthesis 97, 100, 114
C4 plants 99, 115, 246
calorie intake 243
carbohydrate contents 140
carbohydrate synthesis 137
carbohydrates 147
carbon flow into seed starch 143–144
β-carotene 210–211, 216, 219–232
carotenoids 119, 210–211, 228
cattle industry 240–241
cDNA 235
cDNA libraries 234
Celera 39
Centro Internacional de Mejoramiento de
  Maiz y Trigo (CIMMYT) 62, 100–101
centromeres 23, 26–27
cereal crops species genomes 48

254

H

harvest index 117, 136
heat shock tolerance 186
heavy metals 216
Hessian fly 204
*Hind*III fragments 23
*Hind*III library 17, 26
homologous BAC ends 22
HTML 78
hunger, causes of 247–248
Hydrilla 114, 115
Hydrilla verticillata 102, 103, 109
hygromycin 231
hypersensitive response (HR) 154, 163,
    165–166, 196–197

I

ICIS (IRIS) 75
identity preservation systems 241
IMAGE program 15
Impatiens 234
INE (INtegrated rice genome Exploler) 34
Information Network for Genetic Resources
    (SINGER) 63
information resource needs and technologies
    66–76
Institute for Genomic Research (TIGR) rice
    gene index 65
Integrated Genomic Database (IGD) 78
intellectual property rights (IPR) 229
international collaboration 36
International Crop Information System
    (ICIS) 61, 74
International Plant Genetic Resources
    Institute (IPGRI) 63
International Rice Biotechnology Program
    46
International Rice Genebank Collection
    Information System (IRGCIS) 63
International Rice Genome Sequencing
    Project 36, 245
International Rice Information System 74
International Rice Research Institute (IRRI)
    62, 81
iodine deficiency 247
IR64 mutant collection 195–196, 202
IR68 196
IR72 127, 131, 132
IRBB21 194

iron 206–208, 210, 215
iron deficiency 215
iron–zinc interactions 207
irradiance 130
irradiation 194
irrigated environments 7, 9

J

jasmonic acid (JA) 154, 159

K

kinase domains 191
Korean Rice Genome Program 64
Kranz anatomy 102
Kranz anatomy-deficient mutants 97, 114
Kranz leaf anatomy 103
KSU Defense Gene Collection (KSU-DGC)
    197

L

laboratory information management systems
    (LIMS) 61
leaf area index (LAI) 117, 126–127, 132
leaf nitrogen content 132
leaf protein content 126
legumes 58
lesion mimic mutations 196–197, 202
leucine-rich repeats (LRRs) 194
leucine zippers 191
ligand uptake and movement in plant leaves
    91–92
light adaptation 117–134
light-harvesting complexes (LHCs) 121
    LHCI 123
    LHCII 121, 123
light-harvesting proteins 133
light intensity 131
Lipid 162
loss-of-resistance mutants 195
luciferase assays 88
luciferase bioluminescence in intact plants
    92–94
luciferase induction, time course 90
luciferase luminescence 88
lycopene β-cyclase 223, 227
lysine 236, 240, 249
lysine ketoreductase gene 236